Lecture Notes in Control and Information Sciences

Edited by A. V. Balakrishnan and M. Thoma

Vol. 1: Distributed Parameter Systems: Modelling and Identification
Proceedings of the IFIP Working Conference, Rome, Italy, June 21–26, 1976
Edited by A. Ruberti
V, 458 pages. 1978

Vol. 2: New Trends in Systems Analysis
International Symposium, Versailles, December 13–17, 1976
Edited by A. Bensoussan and J. L. Lions
VII, 759 pages. 1977

Vol. 3: Differential Games and Applications
Proceedings of a Workshop, Enschede, Netherlands, March 16–25, 1977
Edited by P. Hagedorn, H. W. Knobloch, and G. J. Olsder
XII, 236 pages. 1977

Vol. 4: M. A. Crane, A. J. Lemoine
An Introduction to the Regenerative Method for Simulation Analysis
VII, 111 pages. 1977

Vol. 5: David J. Clements, Brian D. O. Anderson
Singular Optimal Control: The Linear Quadratic Problem
V, 93 pages. 1978

Vol. 6: Optimization Techniques
Proceedings of the 8th IFIP Conference on Optimization Techniques, Würzburg, September 5–9, 1977
Part 1
Edited by J. Stoer
XIII, 528 pages. 1978

Vol. 7: Optimization Techniques
Proceedings of the 8th IFIP Conference on Optimization Techniques, Würzburg, September 5–9, 1977
Part 2
Edited by J. Stoer
XIII, 512 pages. 1978

Vol. 8: R. F. Curtain, A. J. Pritchard
Infinite Dimensional Linear Systems Theory
VII, 298 pages. 1978

Vol. 9: Y. M. El-Fattah, C. Foulard
Learning Systems:
Decision, Simulation, and Control
VII, 119 pages. 1978

Vol. 10: J. M. Maciejowski
The Modelling of Systems with Small Observation Sets
VII, 241 pages. 1978

Vol. 11: Y. Sewaragi, T. Soeda, S. Omatu
Modelling, Estimation, and Their Applications for Distributed Parameter Systems
VI, 269 pages. 1978

Vol. 12: I. Postlethwaite, A. G. J. McFarlane
A Complex Variable Approach to the Analysis of Linear Multivariable Feedback Systems
IV, 177 pages. 1979

Vol. 13: E. D. Sontag
Polynomial Response Maps
VIII, 168 pages. 1979

Vol. 14: International Symposium on Systems Optimization and Analysis
Rocquentcourt, December 11–13, 1978;
IRIA LABORIA
Edited by A. Bensoussan and J. Lions
VIII, 332 pages. 1979

Vol. 15: Semi-Infinite Programming
Proceedings of a Workshop, Bad Honnef, August 30 – September 1, 1978
V, 180 pages. 1979

Vol. 16: Stochastic Control Theory and Stochastic Differential Systems
Proceedings of a Workshop of the „Sonderforschungsbereich 72 der Deutschen Forschungsgemeinschaft an der Universität Bonn"
which took place in January 1979 at Bad Honnef
VIII, 615 pages. 1979

Vol. 17: O. I. Franksen, P. Falster, F. J. Evans
Qualitative Aspects of Large Scale Systems
Developing Design Rules Using APL
XII, 119 pages. 1979

Vol. 18: Modelling and Optimization of Complex Systems
Proceedings of the IFIP-TC 7 Working Conference
Novosibirsk, USSR, 3–9 July, 1978
Edited by G. I. Marchuk
VI, 293 pages. 1979

Vol. 19: Global and Large Scale System Models
Proceedings of the Center for Advanced Studies (CAS) International Summer Seminar
Dubrovnik, Yugoslavia, August 21–26, 1978
Edited by B. Lazarević
VIII, 232 pages. 1979

Vol. 20: B. Egardt
Stability of Adaptive Controllers
V, 158 pages. 1979

Vol. 21: Martin B. Zarrop
Optimal Experiment Design for Dynamic System Identification
X, 197 pages. 1979

For further listing of published volumes please turn over to inside of back cover.

Lecture Notes in Control and Information Sciences

Edited by A.V. Balakrishnan and M. Thoma

34

H.W. Knobloch

Higher Order Necessary Conditions in Optimal Control Theory

Springer-Verlag
Berlin Heidelberg GmbH 1981

Author
Prof. Dr. H.W. Knobloch
Mathematisches Institut
Am Hubland
D-8700 Würzburg

ISBN 978-3-540-10985-3 ISBN 978-3-540-38548-6 (eBook)
DOI 10.1007/978-3-540-38548-6

Originally published by Springer-Verlag Berlin Heidelberg New York in 1981

2061/3020-543210

PREFACE

The name "higher order conditions" refers to a variety of local tests for optimality which have been introduced into control theory within the last twenty years. These conditions are mostly used as negative criteria in order to rule out the occurence of singular arcs (i.e. portions of optimal solutions for which the bang-bang principle does not hold). The term "higher order" indicates their role as supplement to the standard first order conditions from calculus of variations. In particular they have turned out to be helpful in situations where the maximum principle failed to distinguish between maximizing and minimizing solutions.

These lecture notes are written in the style of a monograph and develop a complete theory of higher order conditions. They are aimed at research workers and users of optimal control methods as well as advanced graduate students who wish to specialize in the field. Some basic knowledge is assumed in optimization and a certain familiarity with fairly elaborate techniques from linear algebra and calculus is desirable.

The approach is straightforward and self-contained, the basic pattern however has its origin in the work of Krener and contributions on the applied side, notably by Jacobson and his co-workers. The author also feels indebted to H.Hermes who drew his attention to the subject and to its differential geometric aspect.

Some remarks seem to be in order concerning general higher order optimization. The method employed here is taylored for control problems described by ordinary differential equations and does not lend itself to generalizations. On the other hand the strength of our result stems from the exploitation of the specific nature of the problem. The crucial point is not the formulation of higher order conditions. It is essential to investigate which conditions are independent and this leads necessarily into algebraic systems theory.

The author thanks Mrs. Ingrid Böhm for typing the manuscript. Her willing and expert participation without which this work would not have been completed is gratefully acknowledged.

Contents.

1. Introduction.

In this work control systems will be considered which are described by ordinary differential equations

$$\dot{x} = f(t,x;u) \quad . \tag{1.1}$$

We assume that f is of class C^∞ with respect to all variables and allow specialization of the control variable u by piecewise C^∞-functions of t which have values in a given set U. Occasionally initial and terminal constraints are imposed on $x(\cdot)$ and the solutions are supposed to minimize an integral of the form

$$\int_{t_o}^{t_e} f^o(t,x;u)dt. \tag{1.2}$$

These solutions are then called optimal.

Higher order necessary conditions are necessary conditions which have to hold along an optimal solution in addition to the Pontryagin maximum principle. They appeared first in connection with the so called singular extremals (see e.g. [8]). Typically, for a singular arc the maximum principle amounts to a set of equality type relations ("first" order conditions), which do not distinguish between arcs maximizing and those minimizing the integral (1.2). This dilemma stimulated the search for a "second order" inequality test which could play a role similar to the standard second order test in calculus. As a result of a combined effort both from the pure and applied side three types of tests which supplement the first order conditions for singular arcs have emerged within the last twenty years and have been successfully applied in the study of concrete examples. They are:

(i) A second order inequality type condition, the so called generalized Clebsch-Legendre condition,

(ii) Second order equality type conditions (conditions of Robbins and Goh) for totally singular arcs (these conditions are meaningful only in case of a multivariable control u),

(iii) Jacobson's condition for free-endpoint problems.

All these conditions are local in nature (i.e. they are expressed in terms of equalities and inequalities which have to hold pointwise along a singular arc) and do not include explicit statements say about conjugate points. There is however a fundamental difference between the first two and the third condition. (i) and (ii) are special cases of a multiplier rule which is described below and in fact is nothing else than a refinement of the maximum principle. It states that a certain time dependent linear homogeneous form is nonnegative on a subset of the state space which can be associated with each point of a prospective singular arc. The coefficients of this linear form are the components of the usual adjoint state vector. Jacobson's condition on the other hand is expressed in terms of a quadratic form with time - varying coefficients. The latter ones depend upon the adjoint state variable <u>and</u> a symmetric time dependent matrix which is solution of a linear matrix differential equation. Jacobson's condition and its generalizations will be treated in part III of this paper whereas part I and II deal with the multiplier rule and its consequences. Altogether the three parts provide a self-contained and unified approach to the analysis of singular extremals via second order conditions. With respect to criteria of type (ii) and (iii) the paper contains results which seem to have so far no analogue in the literature. With respect to known results the paper presents a new angle which may become of interest both from the theoretical and from the computational viewpoint. We mention two aspects which mark the difference of the approach employed in this paper compared with the usual ones. In stating our results no specific assumption about the right hand side of (1.1) is required - except differentiability of sufficient high order (that we assume class C^∞ for f is merely a matter of avoiding too lengthy formulations). Secondly the notion of "singular extremal" will never appear in the subsequent considerations. Instead we will speak of interior controls (part II). These are control functions (or restrictions of those functions to time-intervals) which satisfy the condition $u(t) \in \text{int} U$.

It is in fact only this property of the underlying control function which is needed in order to state all the necessary conditions in a rather satisfactory way. If one adds then the usual hypotheses (normality, linear dependence from the control variable u, etc.) one immediately arrives on the results familiar from the literature.

We now wish to outline very briefly the general lines along which this work is organized. The main result of part I is a multiplier rule (Theorem 9.1) which is developed in Sec. 9-11. These sections can be read independently from the previous ones and are motivated by the paper [3] of Krener. We introduce the notion of a "control variation concentrated at some point $\tilde{t}$". Given a fixed control function $u(\cdot)$, we consider a family $u(\cdot,\lambda)$ of control functions depending on a positive parameter λ such that

$$u(t,\lambda) = u(t) \quad \text{if} \quad t \notin I_\lambda$$

where I_λ is an interval which shrinks to the point $\tilde{t}$ if $\lambda \to 0$. Given furthermore a trajectory $x(\cdot)$ belonging to $u(\cdot)$ one can then associate with $u(\cdot,\lambda)$ a family of trajectories $x(\cdot,\lambda)$ such that

$$x(t,\lambda) \to x(t) \quad \text{for} \quad \lambda \to 0 .$$

Assume now that for $t=\tilde{t}$ the last relation can be strengthened to

$$(1.3) \qquad x(\tilde{t},\lambda) = x(\tilde{t}) + \lambda^s p + o(\lambda^s) \quad \text{as } \lambda \to 0 .$$

for some positive integer s. All elements p of the state space which appear as leading coefficient in an asymptotic expansion (1.3) generated by some control variation concentrated at $\tilde{t}$ constitute a cone in the state space which is denoted by $\Pi^*_{\tilde{t}}$. Next we introduce a further set $\Pi_{\tilde{t}}$, it is - loosely speaking - the set of accumulation points of elements from neighboring sets Π^*_t. We then are in the position to state in a straightforward and geometric way the "higher order necessary conditions". They comprise the maximum principle and the high order maximal principle of Krener [3] and can be phrased as follows:

If $(u(\cdot),x(\cdot))$ is an optimal solution then - in an appropriate augmented state space - Π_t is situated always on one side of a hyperplane which is normal to a certain adjoint state vector. Since the definition of the set Π_t does not offer itself as an effective tool in order to deduce non-trivial conditions we will actually work throughout this paper with a subset of Π_t. In order to define this subset we introduce in Sec. 8 a further subset of the state space which is denoted by $\mathcal{P}_U(t,x,\mathcal{U})$. The definition resembles the definition of Π_t^* since $\mathcal{P}$ is also characterized as collection of leading coefficients in some formal power series which are similar to (1.3). These power series however are not associated with solutions of (1.1). As indicated by the notation the set $\mathcal{P}$ depends upon a triple $t,x,\mathcal{U}$, (t,x) being an arbitrary point in the time-state-space and $\mathcal{U}$ an arbitrary sequence of the form

(1.4) $\{u_o,u_1,\ldots\}$

where each u_i has the same dimension as the control variable u. A more detailed description of the construction which underlies the definition of $\mathcal{P}$ goes like this: One has to specialize, in all possible ways and in terms of functions of a single variable λ, the unknowns and the parameters in a certain type of formal power series which are introduced in the beginning of Sec. 8. There the unknowns are denoted by $z_1,\ldots,z_N$, the parameters which appear in the coefficients $K_{\underline{\nu}}$ of the series are denoted by $\mathcal{U}_o,\ldots,\mathcal{U}_N$, each $\mathcal{U}_i$ being a sequence of the form (1.4). The preceding sections 1-7 of the paper are devoted to a closer study of the $K_{\underline{\nu}}$. It is shown that they can be computed from the right hand side of (1.1) in a purely algebraic fashion. On the other hand there is a more analytic definiton of the $K_{\underline{\nu}}$ which motivates the role which they play in the context of this work. In fact each $K_{\underline{\nu}}$ can be identified with a higher order derivative of the terminal point of a certain trajectory depending upon finitely many parameters. The parameters have the meaning of "variable switching points", i.e. they correspond to points on the t-axis where the definition of the control is switched from one fixed admissible control function to another one. The algorithm for the computation of the $K_{\underline{\nu}}$ hence may be viewed upon as a way of formalizing the well known technique of using "pulse" or "bundle" variations in order to derive necessary conditions for optimal solutions.

In part II of the paper we turn to the problem of constructing elements in Π_t which give rise to conditions other than those which follow from the maximum principle. To this purpose we first associate with a given solution of (1.1) a linear subspace $\mathcal{L}(t)$ of the state space which depends upon t (and of course upon the underlying reference solution). There is an interpretation of $\mathcal{L}(t)$ in terms of notions familiar from linear systems theory ("controllable subspace") which is explained in the beginning of part II (see the exposition preceding Sec. 12). Within the framework of our approach the linear space comes into play if one considers control variations of the form

$$(1.5) \quad u(t) + \lambda^r v(t), \quad v(t) = 0 \quad \text{if} \quad t \notin I_\lambda$$

where I_λ shrinks to $\tilde{t}$ if $\lambda \to 0$. Clearly this variation is concentrated at $\tilde{t}$. Under suitable conditions on $v(t)$ the corresponding trajectory admits an asymptotic expansion of the form

$$(1.6) \quad x(\tilde{t},\lambda) = x(\tilde{t}) + \lambda^r \sum_{\nu=1}^{\infty} p_\nu \lambda^\nu$$

and we have $p_\nu \in \mathcal{L}(\tilde{t})$, $\nu = 1,\ldots,r$. This holds regardless of the special form of $v(\cdot)$. If however $v(t)$ is chosen in such a way that we have

$$(1.7) \quad p_\nu = 0 \quad \text{if} \quad \nu = 1,\ldots, r$$

and if we pick the first non-vanishing among the p_ν with $\nu=r+1,\ldots,2r$ (if there is any) then one obtains an element of $\Pi^*_{\tilde{t}}$ which gives rise to a second order condition. Conversely, all of the second order conditions quoted above can be brought out with the help of a control variation of the form (1.5) which is such that (1.6) and (1.7) hold true. Though this seems to be a very natural access to the second-order theory there remains a serious problem, namely to single out among these conditions the independent ones. We will solve this problem by constructing a basis for the convex cone generated by the relevant coefficients (i.e. the p_ν with $\nu=r+1,\ldots,2r$) in all possible power series (1.6) which arise form control variations of the type (1.5). In case of a scalar u a minimal set of generators for this cone can be specified explicitly as follows. Let n be the dimension of the state variable x and let there be given a fixed reference solution $(u(\cdot),x(\cdot))$ satisfying the condition $u(t) \in \operatorname{int} U$ for all t.

We put

(1.8) $\tilde{f}(t,x) := f(t;x,u(t))$, $B_o(t,x) := (\partial f/\partial u)(t,x;u(t))$

and define a sequence of n-dimensional column vectors depending upon t,x. In explicit terms the recursive relation runs as follows

(1.9) $B_{i+1}(t,x) = (\partial B_i/\partial t)(t,x) + [f,B_i](t,x)$, $i=0,1,\ldots,$

where $[..,..]$ denotes the Lie-bracket of two n-dimensional vectors which depend upon the state variable x. The answer to the question raised above can then be summarized in the following two statements where the symbol p_ν always refers to the coefficient of $\lambda^{\nu+r}$ in some power series (1.6) associated with a control variation of the form (1.5). (i) The vectors $B_i(\tilde{t},x(\tilde{t}))$, $i=0,1,\ldots,$ generate the linear space $\mathcal{L}(\tilde{t})$, i.e. each p_ν with $1 \le \nu \le r$ is a linear combination of $B_i(\tilde{t},x(\tilde{t}))$. (ii) Each p_ν, $r+1 \le \nu \le 2r$, equals an element of the state space which can be constructed in the following way. Starting from these vectors

(1.10) $(\partial^2 f/\partial u^2)(t,x;u(t))$, $[B_{i-1},B_i](t,x)$, $i=1,2,\ldots$

one carries out three kinds of operations, namely:

Repeated application of the operator Γ (see below),
forming linear combinations,
substituting $t \to \tilde{t}$, $x \to x(\tilde{t})$ and adding an element of $\mathcal{L}(\tilde{t})$.

The precise definition of the operator Γ will be given in Sec. 13, for the time being we remark that the application of Γ to a (t,x)-dependent n-dimensional vector $g(t,x)$ results in the vector (cf.(1.8))

$$\frac{\partial g}{\partial t}(t,x) + [\tilde{f},g](t,x)$$

The role of the vectors (1.10) will come out of the considerations in Sec. 12-17, where a certain amount of non-linear systems theory is developed. The approach is straightforward and does not use any advanced Lie-algebra theory, instead we pursue the study of the formal power series and their coefficients $\underline{K}_\nu$ mentioned before. In this way one is led quite naturally to the definition of certain matrices B_i, $i=0,1,\ldots$ (Sec. 13) and to the study of the relations between them. These relations can all be deduced from a fundamental one (Theorem 17.2) which is of interest in its own right. It amounts to an explicit multiplication table for the Lie-brackets $[B_i,B_j]$.

The relevance of this result rests in the fact that the quantities B_i are related to notions which have a well-established meaning both in linear and non-linear control theory. The definition is similar to (1.9) except for the fact that u is not regarded as a given control function u(t) but as an independent variable. In order to make (1.9) then meaningful the B_i have to be regarded as functions of t,x and the further variable $\mathcal{U}$ (cf. (1.4)). If the components of $\mathcal{U}$ are replaced by the successive derivatives of a given control function at a given t one then arrives at the sequence $B_i(t,x)$ which was introduced above. In particular let us consider the case that the underlying system (1.1) is autonomous and of the form

$$\dot{x} = f_o(x) + ug(x),$$

u being a scalar control variable. If $\mathcal{U}$ is then specialized to the sequence {0,0,...} the vector B_i becomes a function of x which is commonly written as $ad^i(f_o)g$ (cf.[3],[21]).

On the other hand if the system is linear then the B_i are independent from $\mathcal{U}$ and are nothing else than the generators of the controllable subspace, in the time-invariant case they are the columns of the Kalman controllability matrix. Hence one may look on the considerations in Sec. 13 as an attempt to define an analogue to this matrix for the general time variant and nonlinear case. So far there is at least one reason which seems to justify this attempt: As a consequence of our definition the fundamental formula for the Lie-brackets $[B_i, B_j]$ (cf. Theorem 17.2) can be written in a rather transparent form. This is not only satisfactory from an aesthetical viewpoint but is also of considerable importance if it comes to the problem of stating the higher order conditions in an applicable form without the usage of ambiguous notation as

$$\frac{\partial}{\partial u} \frac{d^\nu}{dt^\nu} \frac{\partial}{\partial u} H$$

and without imposing restrictions on the right hand side of (1.1) which are needed merely for the purpose of employing a particular mathematical technique. Sec. 20-23 are devoted to the solution of this problem. With respect to the generalized Clebsch-Legendre condition we present two equivalent versions. Theorem 20.1 states this condition in the familiar form; in the case of a multivariable control the assertion is slightly sharper than what can be found in the literature. More interesting is the second version (Theorem 20.2) which seems to be new and particularly suited for applications since one has to compute only half of the number of B_i in order to perform the

the same type of test as with the help of Theorem 20.1. This will be demonstrated by means of a classical example (Lawden's spiral). Sec. 21 deals with equality type conditions and is organized in a similar way as Sec. 20. But besides presenting alternative versions of results stated in a different form elsewhere (cf. [3]) we also establish an entirely new necessary condition for doubly singular arcs (Theorem 21.1). This condition is of interest in situations where f depends in a linear way from one and in a non-linear way from another control variable. Again, Lawden's spiral is an instructive example.

For the benefit of the reader mainly interested in applications some instructions how to read the last sections of part II without going through all of the previous material have been placed in the beginning of part II (before Sec. 12).

Part III of this work has been motivated by Chapter 3 in [7]. Our main result (Theorem 23.1) contains Jacobson's condition as a special case and exhibits the formal structure and the linkage to the generalized Clebsch Legendre condition of this type of second order condition.

Some remarks concerning notation seem to be in order. In general matrix notation will be used, matrices of one column will be called vectors and mostly denoted by small latin letters with $\|..\|$ signifying the Euclidean norm. The transpose of a matrix A is denoted by A^T.

In using the words "variables" and "functions" the additional term "vector-valued" is omitted whenever it is clear from the context that it is meant this way and if there can be no doubts about the dimensions of the vectors. We say that a function is of class C^∞ on some set M if it can be extended to a C^∞ function into some open neighborhood of M. The Jacobian matrix of a function f with respect to a variable, say $x=(x^1,\ldots,x^n)$ (i.e. the matrix consisting of the columns f_{x^i}), is denoted by

$$f_x \quad \text{or} \quad \partial f/\partial x.$$

t always denotes the standard scalar variable "time" and differentation with respect to t is indicated by a dot. In the first part of the paper we frequently have to deal with general solutions

$x(t;t_o,a)$ of differential equations. These are uniquely defined by the following two properties: (i) Regarded as functions of t they are solutions. (ii) We have $x(t_o;t_o,a) = a$ identically in (t_o,a).

We remind the reader of the identity

$$x(t;t_1,x(t_1;t_o,a)) = x(t;t_o,a)$$

which will be constantly used during the manipulations carried out in Section 4.

A special notation has been adopted in order to avoid confusion about the order in which differentiation and substitution of variables has to be performed. It is best explained through the formula (3.2) where this notation is used for the first time. The vertical bar indicates that one should f i r s t perform the differentiations which appear to the left of the bar and t h e n replace w by $w(t)$.

Unless otherwise specified subsets M of the x-space or time-intervals are supposed to be non-empty and open. The closure of M is denoted by $\bar{M}$.

PART I: THE FORMAL FRAMEWORK.

2. Variable Switching Points. The Basic Definitions.

We first introduce some of the standard notation which will be used throughout this paper. t denotes a scalar variable, $\cdot$ means differentiation with respect to t, $x = (x^1,\ldots,x^n)^T$, $f = (f^1,\ldots,f^n)^T$ are n-dimensional column vectors. Without stating this explicitely, we will always assume that a function $f(t,x) = (f^1(t,x),\ldots,f^n(t,x))^T$ is defined and has continuous partial derivatives of any order with respect to t and the components of x on an open set X in (t,x)-space.

Let N be a positive integer and $\tilde{t}$ a real number. We introduce the set

(2.1) $W = W(\tilde{t},N) = \{w = (w^1,\ldots,w^N): w^1 \le w^2 \le \ldots \le w^N \le \tilde{t}\}$.

Throughout the following considerations w will mostly be restricted to a neighbourhood of the corner of W which is farthest to the right, that is the point

(2.2) $\tilde{w} = (\tilde{t},\tilde{t},\ldots,\tilde{t}) = w(\tilde{t})$.

By $w(\tau)$ we will always denote the N-tupel with all components equal to τ .

Assume now that there are given $N + 1$ functions $f_i(t,x)$, $i = 0,\ldots,N$, which are defined on X and which have the differentiability properties mentioned above. We define a function $f(t,x \mid w)$

as follows

$$(2.3)\quad f(t,x\mid w)=\begin{cases} f_o(t,x) & \text{if}\quad t\le w^1\ ,\\ f_i(t,x) & \text{if}\quad w^i<t\le w^{i+1}, i=1,\dots,N-1,\\ f_N(t,x) & \text{if}\quad t> w^N.\end{cases}$$

Except for the points on the hyperplanes $t = w^i$, $i = 1,\dots,N$, $f(t,x\mid w)$ is - for fixed w - a function of class C^∞ on the set X. Furthermore assume that we have a fixed solution $\tilde{x}(t)$ of the diff. eq. $\dot{x} = f_o(t,x)$, which exists on some interval $[t_o,\tilde{t}]$ and satisfies $(t,\tilde{x}(t)) \in X$. Since $f_o(t,x) = f(t,x\mid\tilde{w})$ (according to (2.2) and (2.3)), it follows then by standard arguments that the initial value problem

$$(2.4)\qquad \dot{x} = f(t,x\mid w),\quad x(t_o) = a$$

has a solution $x_w(t;a)$ which also exists on $[t_o,\tilde{t}]$ if $\|w-\tilde{w}\|$ and $\|a-\tilde{a}\|$ are sufficiently small, where $\tilde{a} = \tilde{x}(t_o)$. Furthermore this solution tends to $\tilde{x}(t) = x_{\tilde{w}}(t;\tilde{a})$ for $w \to \tilde{w}$ and $a \to \tilde{a}$. Hence the function

$$(2.5)\qquad \hat{x}(w,a) = x_w(\tilde{t};a)$$

is defined on a set of the form $(\mathcal{N}\cap W)\times\mathcal{A}$, where $\mathcal{N}$ is a neighborhood of $\tilde{w}$ (in the R^N) and $\mathcal{A}$ is a neighborhood of $\tilde{a}$ (in the R^n). In this section we wish to study the analytic properties of the function $\hat{x}(w,a)$.

<u>Lemma 2.1</u> If $\mathcal{N}$ and $\mathcal{A}$ are sufficiently small one can extend the definition of $\hat{x}(w,a)$ to the whole neighorhood $\mathcal{N}\times\mathcal{A}$ in such a way that $\hat{x}(w,a)$ becomes a C^∞-function of (w,a).

<u>Proof.</u> We denote by $x_i(t;\sigma,a)$ the solution of the initial value problem

$$(2.6)\qquad \dot{x} = f_i(t,x),\quad x_i(\sigma) = a,\ i = 0,\dots,N.$$

If we say that x_i exists at certain points of the (t,σ,a)-space, we mean that these points belong to the maximal region of definition of x_i. Note that this region is open, that it contains all points of the form (t,t,a) where $(t,a) \in X$, and that x_i is a C^∞ function of all its arguments on this set. Certainly each x_i exists for this choice of the arguments :

$$t = \tilde{t}, \quad \sigma = \tilde{t}, \quad a = \tilde{a} = \tilde{x}(\tilde{t}).$$

Hence one can find a sequence of neighborhoods $\mathcal{K}_i$ of $\tilde{x}(\tilde{t})$ (in x-space), $i = 1, \ldots, N+1$, and a sequence of positive numbers ε_i, $i = 1,\ldots,N$, such that the following is true

$$\text{(2.7)} \quad \begin{array}{l} x_i \text{ exists on the set } \{t,\sigma,a: |t-\tilde{t}| \le \varepsilon_i, |\sigma-\tilde{t}| \le \varepsilon_i, a \in \mathcal{K}_i\} \\ \text{and satisfies } x_i(t;\sigma,a) \in \mathcal{K}_{i+1},\ i = 1,\ldots,N. \end{array}$$

Furthermore it is clear that $x_o(t;\sigma,a)$ exists in a neighborhood of the set $\{t,\sigma,a: t\in[t_o,\tilde{t}], \sigma=t_o, a=\tilde{x}(t_o)\}$ and is continuous. Hence one can find positive numbers ε_o, δ such that the following statement holds

$$\text{(2.7')} \quad \begin{array}{l} x_o \text{ exists on the set } \{t,\sigma,a: t_o \le t \le \tilde{t} + \varepsilon_o, \|a-\tilde{x}(t_o)\| \le \delta, \sigma=t_o\} \\ \text{and satisfies } x_o(t;t_o,a) \in \mathcal{K}_1 \text{ if } |t-\tilde{t}| \le \varepsilon_o . \end{array}$$

We now choose $\varepsilon = \text{Min}(\varepsilon_o, \varepsilon_1,\ldots,\varepsilon_N)$ and restrict for the remaining part of the proof w,t and a to the set

$$\text{(2.8)} \quad \|w-\tilde{w}\| \le \varepsilon, \quad \|a-\tilde{a}\| \le \delta, \quad |t-\tilde{t}| \le \varepsilon .$$

It can be seen easily from the preceding considerations that for this choice of t,w and a the following functions $y_i(t;w,a)$ are well defined and are C^∞-functions of all their arguments:

$$\text{(2.9)} \quad \begin{array}{l} y_o(t;w,a) = x_o(t;t_o,a) , \\ y_i(t;w,a) = x_i(t;w^i,y_{i-1}(w^i;w,a)),\ i=1,\ldots,N . \end{array}$$

One simply can proceed by induction with respect to i. Besides existence of y_i one has to show that $y_i(t;w,a) \in \mathcal{K}_{i+1}$, which follows from (2.7) and from the choice of ε.

It should be noted that $y_i(t;w,a)$ actually depends on the components $w^1,\ldots,w^i$ of w only. Furthermore we have for $i=1,\ldots,N$

(2.10) $\quad y_i(w^i;w,a) = y_{i-1}(w^i;w,a)$ if $w = (w^1,\ldots,w^i,\ldots,w^N)$.

This is a consequence of the identity $x_i(t,t,a) = a$. Putting $w=\tilde{w}$ (cf. (2.2)) we obtain from (2.10) and (2.9).

(2.11) $\quad y_i(\tilde{t};\tilde{w},a) = x_o(\tilde{t};t_o,a),\ i = 0,\ldots,N$.

Assume now, that $w \in \text{int}(W)$ (that is the set of all $w=(w^1,\ldots,w^N)$ such that $w^1 < w^2 < \ldots < w^N < \tilde{t}$) . It is not difficult to verify the correctness of the following statements

$$\begin{aligned} y_o(t;w,a) &= x_w(t;a) \quad \text{for} \quad t \le w^1 , \\ y_i(t;w,a) &= x_w(t;a) \quad \text{for} \quad w^i < t \le w^{i+1},\ i = 1,\ldots,N-1, \\ y_N(t;w,a) &= x_w(t;a) \quad \text{for} \quad t > w^N. \end{aligned} \tag{2.11'}$$

One has to use the fact, that both y_i and x_w , regarded as functions of t, are solutions of the differential eq. $\dot{x} = f_i(t,x)$ for $w^i < t < w^{i+1}$. Furthermore their initial values at $t= w^i$ coincide. This can be established by induction with respect to i using (2.10).

Hence it follows from (2.5) that

$$\begin{aligned} &\hat{x}(w,a) = y_N(\tilde{t};w,a) = x_w(\tilde{t};a) \\ &\text{if} \quad w \in \text{int}(W),\ \ \|w-\tilde{w}\| \le \epsilon,\ \|a-\tilde{a}\| \le \delta . \end{aligned} \tag{2.12}$$

For reasons of continuity (2.12) holds also if $w \in W$. Thereby the lemma is proved, since, as we have noted before, the function on the right hand side of (2.12) is a C^∞-function of (w,a) .

The preceding considerations can be extended to the case of parameter-dependent equations. Let $f_i = f_i(t,x,p)$ be C^∞-functions of t,x,p for $(t,x) \in X$ and $p \in \mathcal{P}_o$, $\mathcal{P}_o$ an open set in p-space. Let $f(t,x,p \mid w)$ be defined as in (2.3) with $f_i(t,x,p)$ instead of $f_i(t,x)$ and denote by $x_w(t;a,p)$ the solution of

(2.13) $\quad \dot{x} = f(t,x,p \mid w)$, $x(t_o) = a$.

Assume that $\tilde{x}(t)$ is solution of $\dot{x} = f_o(t,x,\tilde{p})$. Then $x_w(t;a,p)$ exist on $[t_o,\tilde{t}]$ if $\|w-\tilde{w}\|$, $\|a-\tilde{a}\|$, $\|p-\tilde{p}\|$ are sufficiently small. Furthermore $x_w(\tilde{t};a,p) = \hat{x}(w,a,p)$ can be extended to a C^∞-function on a set of the form $\mathcal{N} \times \mathcal{A} \times \mathcal{P}$, where $\mathcal{N}$, $\mathcal{A}$, $\mathcal{P}$ are full neighborhood of $\tilde{w}, \tilde{a}, \tilde{p}$ respectively. This result follows from the preceding one, since formally the parameters can be considered as initial values. One simply has to replace the initial value problem (2.13) by this one

$$\dot{x} = f(t,x,y|w),\ \dot{y} = 0,\ x(t_o) = a, y(t_o) = p.$$

In what follows D denotes some partial differential operator with respect to the components of w and a. In the remaining part of this section we want to develop some basic relations between the values which the partial derivatives of the function $\hat{x}(w,a)$ assume for $w=\tilde{w}$, $a=\tilde{a}$. This values are denoted by $(D\hat{x})(\tilde{w},\tilde{a})$. In order to simplify our considerations we introduce for the moment the following notation:

(2.14) $\quad w_i(w^1,\ldots,w^i) = (w^1,\ldots,w^i, \tilde{t},\ldots\tilde{t}).$

Hence $w_i(w^1,\ldots,w^i)$ is the N-tupel w with variabel components $w^1,\ldots,w^i$ and fixed $w^j = \tilde{t}$ for $j > i$. Now if one specializes w to $w_i(w^1,\ldots,w^i)$ in the formula (2.10) (with j instead of i) and if $j > i$ then one obtains

$$y_j(\tilde{t};w_i(w^1,\ldots,w^i),a) = y_{j-1}(\tilde{t};w_i(w^1,\ldots,w^i),a) .$$

Repeated application of this formula yields the identity

$$\hat{x}(w_i(w^1,\ldots,w^i),a) = y_N(\tilde{t};w_i(w^1,\ldots,w^i),a) = y_i(\tilde{t};w_i(w^1,\ldots,w^i),a)$$

for $i=1,\ldots,N$. Now y_i depends upon the first i components of w only, as we have remarked before. Hence

$$y_i(\tilde{t};w,a) = y_i(\tilde{t};w_i(w^1,\ldots,w^i),a), \text{ if } w = (w^1,\ldots,w^i,\ldots).$$

It follows therefore from the last relation that

(2.15) $\quad (D\,\hat{x})(\tilde{w},\tilde{a}) = (Dy_N)(\tilde{t};\tilde{w},\tilde{a}) = (Dy_i)(\tilde{t};\tilde{w},\tilde{a})$

in case D is a differential operator which involves no differentiation with respect to all w^j for $j > i$. Especially if D involves differentiation with respect to the components of a only then we obtain from (2.11), (2.12)

$$(2.16) \qquad (D\hat{x})(\tilde{w},\tilde{a}) = (Dy_j)(\tilde{t};\tilde{w},\tilde{a}) = (Dx_o)(\tilde{t};t_o,\tilde{a}),\ j=1,\dots,N.$$

3. The formula for $(D\hat{x})(w,a)$ in case $D = \partial^\nu/(\partial w^i)^\nu$.

In this and the following section we keep the initial value a fixed: $a = \tilde{a} = \tilde{x}(t_o)$. For shortness we write $y_i(t;w)$, $\hat{x}(w)$ instead of $y_i(t;w,\tilde{a})$, $\hat{x}(w,\tilde{a})$. In what follows we make frequently use of the relation (2.11) which in the present notation assumes the form

$$(3.1) \qquad \hat{x}(\tilde{w}) = y_i(\tilde{t};\tilde{w}) = \tilde{x}(\tilde{t}),\ i=0,\ \dots\ ,\ N.$$

Note that $x_o(t;t_o,\tilde{a}) = \tilde{x}(t)$ (= reference trajectory).

Our aim is to construct for every $i=1,\dots,N$ and every $\nu = 0,1,\dots$ a function $h_i^{(\nu)}(t,x)$ of the two variables t and x for which the relation

$$(3.2) \qquad h_i^{(\nu)}(w^i,y_i(w^i;w)) = \frac{\partial^\nu y_i}{(\partial w^i)^\nu}(t;w)\Bigg|_{t\mapsto w^i}$$

holds identically in $w = (w^1,\dots,w^i,\dots,w^N)$. Before we proceed to the construction we note that by putting $w=\tilde{w}$ (c.f. (2.2)) and using (3.1) one obtains from (2.15) and (3.2) the relation

$$(3.3) \qquad h_i^{(\nu)}(\tilde{t},\tilde{x}(\tilde{t})) = \frac{\partial^\nu \hat{x}}{(\partial w^i)^\nu}(\tilde{w}) .$$

This is the formula quoted in the title of this section; it will play an important role in our further considerations.

For notational convenience we introduce the following abbreviation. Let $g(x)$ be a function of x (infinitely often differentiable), where g and x are n-dimensional vectors. We then use the symbol

$$(\mathcal{D}^\nu g)(x_o,\dots,x_\nu)$$

to denote a n-dimensional vector-valued function depending upon the n-dimensional vectors $x_o,\dots,x_\nu$, which is defined uniquely by the following property: If $x(\sigma)$ is any (n-dimensional vector -)

function of the scalar variable σ and if $x(\sigma)$ is ν-times differentiable then

$$(3.4)\qquad (\mathcal{D}^{\nu}g)(x(\sigma), \frac{dx}{d\sigma}(\sigma),\ldots,\frac{d^{\nu}x}{d\sigma^{\nu}}(\sigma)) = \frac{d^{\nu}}{d\sigma^{\nu}}\, g(x(\sigma)) \ .$$

In other words: One obtains the ν-th derivative (with respect to σ) of the composite function $g(x(\sigma))$ from $\mathcal{D}^{\nu}g$ by means of the substitution $x_i \to \frac{d^i x}{d\sigma^i}(\sigma)$, $i = 0,\ldots,\nu$. If g depends on variables other than x , say on t, then the symbol

$$(\mathcal{D}^{\nu}_{x}g)(t,x_0,\ldots,x_{\nu})$$

indicates that the operator $\mathcal{D}^{\nu}$ has to be applied with respect to the variable x only. Using (3.4) $\mathcal{D}^{\nu}g$ is easily calculated for $\nu = 0,1,2$:

$$(3.5)\qquad \begin{aligned} &(\mathcal{D}^{0}g)(x_0) = g(x_0), \quad (\mathcal{D}^{1}g)(x_0,x_1) = g_x(x_0)\cdot x_1 \ , \\ &(\mathcal{D}^{2}g)(x_0,x_1,x_2) = (g_x(x_0)x_1)_{x_0}\cdot x_1 + g_x(x_0)\cdot x_2 \ . \end{aligned}$$

Here g_x denotes the functional matrix of g and $(\ldots)_{x_0}$ denotes the functional matrix with respect to x_0 of the vector-valued function $g_x(x_0)\cdot x_1$. - From the relation (3.4) one obtains immediately the following general recursion formula

$$\mathcal{D}^{\nu+1}g = \sum_{\mu=0}^{\nu} (\mathcal{D}^{\nu}g)_{x_{\mu}}\cdot x_{\mu+1} \ .$$

Before we turn to the main subject of this section, we wish to state a lemma, which will be needed later. The proof is a straightforward application of (3.5) and (3.6).

<u>Lemma 3.1</u> We have for $\nu \geq 1$ a representation of the form

$$(\mathcal{D}^{\nu}g)(x_0,x_1,\ldots,x_{\nu}) = g_x(x_0)\cdot x_{\nu} + r_{\nu}(x_0,x_1,\ldots,x_{\nu-1})$$

where the components of the remainder term r_{ν} have the following property. They can be written as polynomials in the components of $x_1,\ldots,x_{\nu-1}$ and vanish together with their first order partial derivatives at $x_1 = x_2 = \dots = x_{\nu-1} = 0$. The coefficients of these polynomials are partial derivatives of the components of $g(x)$ of order $\leq \nu$, evaluated at $x = x_o$. In particular, if the values of two functions $g, \hat{g}$ and the values of their derivatives up to order ν coincide at some particular $x_o = \tilde{x}_o$, then

$$(\mathcal{D}^\nu g)(\tilde{x}_o, x_1, \dots, x_\nu) = (\mathcal{D}^\nu \hat{g})(\tilde{x}_o, x_1, \dots, x_\nu)$$

identically in $x_1, \dots, x_\nu$.

Throughout the following considerations the components of w except w^i have to be regarded as fixed. For notational convenience we use for the moment the symbols in the first line of the list below instead of the corresponding symbols in the second line

ω	$y(t)$	$z(t,\omega)$	$k(t,x)$	$l(t,x)$
w^i	$y_{i-1}(t;w)$	$y_i(t;\omega)$	$f_{i-1}(t,x)$	$f_i(t,x)$

(Note that $y_{i-1}(t;w)$ actually does not depend upon $\omega = w^i$, see the remark following (2.9)). We now write down three relations which are satisfied identically in (t,ω). They are immediate consequences of (2.10) and of what was said about y_i later (cf. (2.11')).

$$\text{(3.7)} \qquad \begin{aligned} &\frac{\partial}{\partial t} z(t,\omega) = l(t, z(t,\omega)), \\ &\dot{y}(t) = k(t, y(t)), \; z(\omega,\omega) = y(\omega). \end{aligned}$$

If the first of these identities is differentiated ν times with respect to ω and if order of differentiation is interchanged we obtain

$$\frac{\partial}{\partial t}\left(\frac{\partial^\nu}{\partial \omega^\nu} z(t,\omega)\right) = (\mathcal{D}_x^\nu l)(t, x_o, \dots, x_\nu)\Big|_{x_j \to \frac{\partial^j z}{(\partial\omega)^j}(t,\omega),\, j=0,\dots,\nu}.$$

Next we introduce the functions

$$z^{(0)}(\omega) := z(\omega,\omega) = y(\omega) \quad (\text{cf. } (3.7))$$

$$z^{(\nu)}(\omega) := \frac{\partial^\nu z}{\partial \omega^\nu}(t,\omega)\Big|_{t\to\omega}, \quad \nu = 1,2,\ldots$$

Using the preceding identity we obtain a recursion formula for $z^{(\nu)}(\omega)$ by differentiation of the defining relation with respect to ω:

$$\frac{dz^{(\nu)}}{d\omega}(\omega) = \frac{\partial}{\partial t}\left(\frac{\partial^\nu z}{\partial \omega^\nu}(t,\omega)\right)\Big|_{t\to\omega} + \frac{\partial^{\nu+1} z}{\partial \omega^{\nu+1}}(t,\omega)\Big|_{t\to\omega} \tag{3.8}$$

$$= (\mathcal{D}_x^\nu 1)(\omega, z^{(o)}(\omega),\ldots,z^{(\nu)}(\omega)) + z^{(\nu+1)}(\omega) .$$

From this recursion formula one sees immediately that $z^{(\nu)}(\omega)$ admits a representation of the form $h^{(\nu)}(\omega,y(\omega))$, where $h^{(\nu)}(t,x)$ is a function of t and x, which can be expressed directly in terms of the functions $l(t,x)$ and $k(t,x)$. This is true for $\nu = 0$, since $z^{(o)}(\omega) = y(\omega)$. Furthermore, if we have a representation

$$z^{(\nu)}(\omega) = h^{(\nu)}(\omega,y(\omega)),$$

and if we use the fact that $y(t)$ satisfies the differential eq. $\dot{y} = k(t,y)$ we obtain

$$\frac{dz^{(\nu)}}{d\omega}(\omega) = h_t^{(\nu)}(\omega,y(\omega)) + (h^{(\nu)})_x(\omega,y(\omega)) \cdot k(\omega,y(\omega)) .$$

Hence we are arrived at the following result. Let the functions $h^{(\nu)}(t,x)$ be defined recursively as follows: $h^{(o)}(t,x) = x$,

$$h^{(\nu+1)}(t,x) = \frac{d}{dt} h^{(\nu)}(t,x) - (\mathcal{D}_x^\nu 1)(t,h^{(o)}(t,x),\ldots,h^{(\nu)}(t,x)),$$

where for the moment $\frac{d}{dt}$ means differentiation with respect to the differential equation $\dot{y} = k(t,y)$. Then

$$z^{(\nu)}(\omega) = h^{(\nu)}(\omega,y(\omega)) . \tag{3.9}$$

We now return to the original notation and obtain from (3.9) the desired representation for the expression on the right hand side of (3.2) in terms of $\omega = w^i$ and $y(\omega) = z(\omega,\omega) = y_i(w^i;w)$.

<u>Theorem 3.1.</u> Let the functions $h_i^{(\nu)}(t,x)$ be defined recursively as follows: $h_i^{(o)}(t,x) = x$,

$$h_i^{(\nu+1)}(t,x) = \frac{\partial}{\partial t} h_i^{(\nu)}(t,x) + \left(h_i^{(\nu)}\right)_x \cdot f_{i-1}(t,x)$$

$$- (\mathcal{D}_x^\nu f_i)(t,h_i^{(o)}(t,x),\ldots,h_i^{(\nu)}(t,x)) ,$$

where $(\mathcal{D}_x^\nu f_i)(t,x_o,\ldots,x_\nu)$ has to be understood according to definition (3.4).

Then the identies (3.2) and (3.3) hold.

Examples. We make use of (3.5) and omit occasionally the argument (t,x)

$$h_i^{(o)}(t,x) = x, h_i^{(1)} = f_{i-1} - f_i ,$$

(3.10)
$$h_i^{(2)} = \left(h_i^{(1)}\right)_t + \left(h_i^{(1)}\right)_x f_{i-1} - (f_i)_x h_i^{(1)}$$

(3.11)
$$= (f_{i-1})_t - (f_i)_t + (f_{i-1})_x f_{i-1} + (f_i)_x f_i - 2(f_i)_x f_{i-1} ,$$

$$h_i^{(3)}(t,x) = (h_i^{(2)})_t(t,x) + (h_i^{(2)})_x(t,x) \cdot f_{i-1}(t,x)$$

$$- \left((f_i)_x(t,y) \cdot h_i^{(1)}(t,x)\right)_y \cdot h_i^{(1)}(t,x)\Big|_{y \to x}$$

$$- (f_i)_x(t,x) h_i^{(2)}(t,x) .$$

Corollary. The components of $h_i^{(\nu)}$ can be written as polynomials in the components of f_i, f_{i-1} and their derivatives (with respect to x and t) up to order $\nu-1$.

Proof. By induction with respect to ν. For $\nu = 1$ the statement is evident in view of (3.10). The conclusion in the general case follows easily from the recursion formula given in the theorem and from Lemma 3.1.

4. Dependence upon the f_i.

In this section we are going to establish some basic facts about the way in which the function $\hat{x}(w)$ depends upon the choice of the functions f_i which enter into the definition of f (c.f. (2.3)). The main result will be a formula given in Theorem 4.2 which resembles somehow the so called Campbell-Hausdorff formula . It will play an essential role in our further considerations. We conclude this section with three corollaries, which are straightforward applications of the theorem. A further application will be given in Section 5 where we derive explicite formulae for $D\hat{x}(w)$ in case D is a differential operator which acts on more than one component of w.

We resume the discussion of the functions $x_i(t;\sigma,a)$ introduced in Lemma 2.1. Let us recall that each function x_i has a certain unique maximal region of definition $\mathcal{D}_i$ which is an open set of points of the form (t,σ,x) where $(t,x)\in X$ (X being the common region of definition of the f_i). Note that $(t,t,x)\in \mathcal{D}_i$ whenever $(t,x)\in X$.

First we introduce a series of auxiliary functions $\hat{y}_i(\tau,t,w,x)$ which depend upon the variables τ,t,w,x. Here τ,t are scalars and x,w respectively denote as before the n-tuple $(x^1,\ldots,x^n)$ and the N-tuple $(w^1,\ldots,w^N)$ respectively. The $\hat{y}_i$ are n-dimensional vectors and are recursively defined as follows

$$(4.1)\qquad \begin{aligned} \hat{y}_o(\tau,t,w,x) &= x_o(\tau;t,x) \ , \\ \hat{y}_i(\tau,t,w,x) &= x_i(\tau;w^i,\hat{y}_{i-1}(w^i,t,w,x)),\ i=1,\ldots,N. \end{aligned}$$

It is immediately clear, that each $\hat{y}_i$ is of class C^∞ on an open subset of the (τ,t,w,x)-space, and that this set contains all points of the form $(t,t,w(t),x)$ where $(t,x)\in X$ (for the definition of $w(t)$ cf. (2.2)). It also follows from the definition that $\hat{y}_i$ actually depends upon τ,t,x and the components $w^1,\ldots,w^i$ of w only and that we have

$$(4.2)\qquad \hat{y}_i(t,t,w(t),x) = x.$$

We are particularly interested in $\hat{y}_N(t,t,w,x)$ and will therefore introduce a special symbol for this function. This symbol will also indicate that the construction of $\hat{y}_N(\ldots)$ depends upon the choice of $f_o,\ldots,f_N$.

<u>Definition 4.1.</u> Given a positive integer N and $N+1$ functions $f_o,\dots,f_N$ which are defined and of class C^∞ on some set X in (t,x)-space. Let $\hat{y}_i$ be defined according to (4.1), where $x_i(t;\sigma,a)$ is the general solution of the diff. eq. $\dot{x} = f_i(t,x)$. The function $\hat{y}_N(t,t,w,x)$ will then be denoted by

$$c(t,x;w\,|\,f_o(\cdot),\dots,f_N(\cdot))$$

or, if necessary, more detailed by

$$(4.3)\qquad c(t,x;w^1,\dots,w^N\,|\,f_o(\cdot),\dots,f_N(\cdot)).$$

The advantage of this notation lies in the fact that one can express certain formal relations in a rather transparent way as we are going to show in the next lemma. It is also because of this notation that the fundamental result of this section - Theorem 4.2 - becomes a convenient and useful tool for our further considerations.

<u>Lemma 4.1.</u> The following relations hold identically in $t,x,w^1,\dots,w^N$.

(i) $\quad c(t,x;t,\dots,t\,|\,f_o(\cdot),f_1(\cdot),\dots,f_N(\cdot)) = x,$

(ii) $\quad c(t,x;w^1,\dots,w^N\,|\,f_o(\cdot),\ f_o(\cdot),\dots,f_o(\cdot)) = x,$

(iii) $\quad c(t,x;w^1,\dots,w^1\,|\,f_o(\cdot),f_1(\cdot),\dots,f_{N-1}(\cdot),f_N(\cdot)) =$
$\quad = c(t,x;w^1\,|\,f_o(\cdot),f_N(\cdot)).$

<u>Proof.</u> It follows from (4.1) by induction with respect to i that

$$\hat{y}_i(\tau,t,w(t),x) = x_i(\tau;t,x)$$

for $i = 0,\dots,N$. For $i = N$ and $\tau = t$ one obtains the first statement of the lemma.

In order to prove (ii), we remark that $f_i = f_o$ for $i = 1,\dots,N$ implies

$$x_i(t;\sigma,a) = x_o(t;\sigma,a)$$

identically in (t,σ,a), $i=1,\dots,N$. Hence, using (4.1) and induction with respect to i, we obtain

$$\hat{y}_i(\tau,t,w,x) = x_o(\tau;t,x)$$

$i = 0,\dots,N$. Taking $i = N$ and $\tau = t$ we arrive at the desired result.

We next turn to (iii) and assume for the rest of the proof that w has the special form

$$w = (w^1, w^1, \dots, w^1).$$

Again, from (4.1), one sees immediately by induction that in this case

$$\hat{y}_i(\tau, t, w, x) = x_i(\tau; w^1, x_o(w^1, t, x))$$

$i = 1, \dots, N$. Hence

$$\hat{y}_N(t, t, w, x) = x_N(t; w^1, x_o(w^1; t, x)),$$

and this is nothing else than what is stated in part (iii) of the lemma. Indeed, for $N=1$ the definition of $c(\dots)$ reduces to the simple form

$$c(t,x;w^1 | f_o(\cdot), f_1(\cdot)) = x_1(t; w^1, x_o(w^1; t, x)) \,. \tag{4.4}$$

The next theorem will provide the link between the definitions introduced in this and in the previous sections.

<u>Theorem 4.1</u> Let there be given an integer N, functions $f_o, \dots, f_N$ and a solution $\tilde{x}(t)$ of $\dot{x} = f_o(t,x)$ on some interval $[t_o, \tilde{t}]$. Let $f(t,x|w)$ and $x_w(t;a)$ be defined as in Section 2. Then

$$x_w(\tilde{t}; \tilde{x}(t_o)) = \hat{x}(w) = c(\tilde{t}, \tilde{x}(\tilde{t}); w | f_o(\cdot), \dots, f_N(\cdot)) \,.$$

<u>Proof.</u> We begin by writing down an identity which follows from the definition of $x_o(t;\sigma,a)$ (note that both $\tilde{x}(t)$ and $x_o(t;\sigma,a)$ are solutions of the diff. eq. $\dot{x} = f_o(t,x)$)

$$x_o(t; \tilde{t}, \tilde{x}(\tilde{t})) = x_o(t; t_o, \tilde{x}(t_o)) = \tilde{x}(t) \,.$$

Performing the substitutions $\tau \to t$, $t \to \tilde{t}$, $x \to \tilde{x}(\tilde{t})$ on both sides of (4.1) and using this identity yields then the relations

$$\hat{y}_o(t, \tilde{t}, w, \tilde{x}(\tilde{t})) = x_o(t; t_o, \tilde{x}(t_o)),$$
$$\hat{y}_i(t, \tilde{t}, w, \tilde{x}(\tilde{t})) = x_i(t; w^i, \hat{y}_{i-1}(w^i, \tilde{t}, w, \tilde{x}(\tilde{t}))).$$

This is exactly the same recursive scheme (2.9) which was used previously in order to define the y_i. Hence we have

$$\hat{y}_i(t,\tilde{t},w,\tilde{x}(\tilde{t})) = y_i(t;w,\tilde{x}(t_o))$$

for $i=0,\dots,N$. Taking $i=N$ and $t=\tilde{t}$ we obtain then the desired result, in view of (2.12) and Definition 4.1.

We now are in a position to present the central result of this section.

Theorem 4.2.

Let M,N be positive integers such that $1 \le M < N$. Let there be given, as before, $N+1$ functions f_i, $i=0,\dots,N$, which have a common region of definition X. Furthermore let $(\tilde{t},\tilde{x})$ be an arbitrary point of X and let p be the triple consisting of t,x and $w(t)$ (cf. (2.2)).

Then there exists a neighborhood $\mathcal{N}$ of p in (t,x,w)-space such that the following relation holds identically in t,x,w on $\mathcal{N}$

$$\begin{aligned} c(t,x;w^1,\dots,w^N|f_o(\cdot),\dots,f_N(\cdot)) &= \\ = c(t,x';w^{M+1},\dots,w^N|f_M(\cdot),\dots,f_N(\cdot)) \end{aligned} \tag{4.5}$$

where

$$x' = c(t,x;w^1,\dots,w^M|f_o(\cdot),\dots,f_M(\cdot)). \tag{4.6}$$

Proof. We first note that all expressions occuring in (4.5), (4.6) are well defined if $t=\tilde{t}$ $x=\tilde{x}$, $w^i=\tilde{t}$ for $i=1,\dots,N$, and they assume the same value $\tilde{x}$. This follows from part (i) of Lemma 4.1. Since all functions involved are continuous (in t,x,w) and have an open region of definition it is clear that one can find a neighborhood $\mathcal{N}$ of p such that the left hand side of (4.5) and the right hand sides of (4.5) and (4.6) are meaningful, if $(t,x,w) \in \mathcal{N}$.

We have, according to Definition 4.1,

$$x' = c(t,x;w^1,\dots,w^M|f_o(\cdot),\dots,f_M(\cdot)) = \hat{y}_M(t,t,w,x), \tag{4.7}$$

where $\hat{y}_M$ is the M+1-th member of the recursive system (4.1).

Note that $\hat{y}_M$ depends upon the components $w^1 \ldots, w^M$ of w only.

In order to construct the expression on the right hand side of (4.5) let us consider the functions $\tilde{y}_i(\tau,t,w,x)$ which are recursively defined as follows

$$\begin{aligned} \tilde{y}_M(\tau,t,w,x) &= x_M(\tau;t,x), \\ \tilde{y}_i(\tau,t,w,x) &= x_i(\tau;w^i,\tilde{y}_{i-1}(w^i,t,w,x)),\ i=M+1,\ldots,N. \end{aligned} \tag{4.8}$$

We have then, also according to Definition 4.1,

$$c(t,x;w^{M+1},\ldots,w^N|f_M(\cdot),\ldots,f_N(\cdot)) = \tilde{y}_N(t,t,w,x). \tag{4.9}$$

We now claim that the following relations hold

$$\tilde{y}_M(w^{M+1},t,w,x') = \hat{y}_M(w^{M+1},\ t,w,x) \tag{4.10}$$

identically in w,t,x, where x' is given by (4.7),

$$\tilde{y}_i(\tau,t,w,x') = \hat{y}_i(\tau,t,w,x), i=M+1,\ldots,N, \tag{4.11}$$

identically in τ,t,w,x, where x' is given by (4.7).

Comparing the recursion formulae (4.1) and (4.8) one sees immediately hat (4.11) is a consequence of (4.10). So all we need to do is to prove (4.10). From (4.1) and (4.7) we obtain

$$x' = x_M(t;w^M,\hat{y}_{M-1}(w^M,t,w,x)).$$

Substituting x' for x into the formula for $\tilde{y}_M$ (cf. (4.8)) and using the above representation of x' we arrive at this result

$$\tilde{y}_M(w^{M+1},t,w,x') = x_M(w^{M+1};w^M,\hat{y}_{M-1}(w^M,t,w,x)),$$

which turns out to be nothing else than (4.10) (take $i=M$ in (4.1)!).

The proof of the theorem is now easily completed. Since $M < N$ the statement (4.11) is true in particular for $i=N$ and $\tau = t$. But in this case it can be written also in the form (4.5), as can be seen immediately from (4.9) and from Definition 4.1.

Corollary 1. If $f_i = f_o$ for $i=1,\dots,M$, then

$$c(t,x;w^1,\dots,w^N|f_o(\cdot),\dots,f_N(\cdot)) = \\ = c(t,x;w^{M+1},\dots,w^N|f_M(\cdot),\dots,f_N(\cdot))$$

identically in t,x,w.

Proof. Follows immediately from Theorem 4.2 and from Lemma 4.1, part (ii).

Corollary 2. Let D be a differential operator, which does not act on the variables $t,x,w^1,\dots,w^M$ (that is, D acts on some of the variables $w^{M+1},\dots,w^N$ only). Then the relation

$$(Dc)(t,x;w^1,\dots,w^N|f_o(\cdot),\dots,f_N(\cdot)) = \\ = (Dc)(t,x;w^{M+1},\dots,w^N|f_M(\cdot),\dots,f_N(\cdot))$$

holds on the set

$$(4.12) \quad \{t,x,w = (w^1,\dots,w^N):w^i = t \quad \text{for} \quad i=1,\dots,M\}.$$

Proof. Since D does not act on $w^1,\dots,w^M$, the function Dc coincides on the set (4.12) with $D\hat{c}$, where $\hat{c}$ is equal to

$$(4.13) \quad c(t,x;t,\dots,t,w^{M+1},\dots,w^N|f_o(\cdot),\dots,f_N(\cdot)).$$

It follows now from Theorem 4.2 and from Lemma 4.1, part (i), that (4.13) can be written as

$$c(t,x';w^{M+1},\dots,w^N|f_M(\cdot),\dots,f_N(\cdot))$$

where

$$x' = c(t,x;t,\dots,t|f_o(\cdot),\dots,f_M(\cdot)) = x.$$

This proves the Corollary.

Corollary 3. Let $i_1,i_2,\dots,i_L$ be integers such that

$$1 \le i_1 < i_2 < \dots < i_L = N.$$

Then the relation

$$(4.14) \quad \begin{aligned} &c(t,x;w^1,\dots,w^N|f_o(\cdot),\dots,f_N(\cdot)) = \\ &= c(t,x;w^{i_1},\dots,w^{i_L}|f_o(\cdot),f_{i_1}(\cdot),\dots,f_{i_L}(\cdot)) \end{aligned}$$

holds on the set

$$(4.15) \quad \{t,x,w:w^1 = w^2 = \dots = w^{i_1}, w^{i_1+1} = \dots = w^{i_2},\dots,w^{i_{L-1}+1} = \dots = w^{i_L}\}.$$

<u>Proof.</u> If $L = 1$ then we have $i_1=N$ and the statement of the corollary is identical with part (iii) of Lemma 4.1. So we proceed by induction with respect to L. Applying Theorem 4.2 (with $M = i_1$) the left hand side of (4.14) can be changed to

(4.16) $$c(t,x';w^{i_1+1},\dots,w^N|f_{i_1}(\cdot),\dots,f_N(\cdot))$$

where $x' = c(t,x;w^1,\dots,w^1|f_o(\cdot),\dots,f_{i_1}(\cdot))$.

Here we have already assumed that (t,x,w) belongs to the set (4.15). Since this assumption implies also that the components of w satisfy the condition

$$w^{i_1+1}=\dots=w^{i_2},w^{i_2+1}=\dots=w^{i_3},\dots,w^{i_{L-1}+1}=\dots=w^{i_L},$$

one can use hypothesis of induction and write (4.16) in the form

(4.17) $$c(t,x';w^{i_2},\dots,w^{i_L}|f_{i_1}(\cdot),\dots,f_{i_L}(\cdot)).$$

On the other hand we have, in view of Lamm 4.1, part (iii),

$$x' = c(t,x;w^1|f_o(\cdot),f_{i_1}(\cdot)).$$

Applying Theorem 4.2 once more (with $M=1$) and using the fact that $w^1 = w^{i_1}$ one sees immediately that (4.17) coincides with the right hand side of (4.14). Thereby the corollary is proved.

<u>5. Differentiation with respect to w.</u>

As we have seen in the previous section the function

$$c(t,x;w^1,\dots,w^N|f_o(\cdot),\dots,f_N(\cdot))$$

is defined and a C^∞-function of (t,x,w) on an open set which contains all points of the form $(t,x,w(t))$ with $(t,x)\in X$. Throughout this section D will denote a differential operator which acts on the components of w. D can be written as a formal product

(5.1) $$D = D_1^{\nu_1}\circ D_2^{\nu_2}\circ\dots\circ D_N^{\nu_N}, \text{ where}$$
$$D_i= \partial/\partial w^i, \quad i = 1,\dots,N ,$$

and $\nu_i \geq 0$ are integers. In this section we introduce certain functions of the variables t,x, which are associated to the differential operators D and which will play an important role in the future. These functions will be denoted by $k_D(t,x)$ or, occasionally, by $k_D(t,x|f_o(\cdot),\dots,f_N(\cdot))$ in order to indicate the dependence from the choice of the functions $f_o,\dots,f_N$. The precise definition is as follows

$$(5.2)\quad k_D(t,x|f_o(\cdot),\dots,f_N(\cdot)) = \\ = (Dc)(t,x;w^1,\dots,w^N|f_o(\cdot),\dots,f_N(\cdot))\Big|_{w\to w(t)} .$$

It is clear, in view of our previous remarks, that k_D is well defined on X and is a C^∞-function of its arguments. There are certain formal relations between the functions k_D which are immediate consequences of the definition (5.2) and of the relations existing between the functions $c(\dots)$. We list two of them which will be needed in the sequel.

Lemma 5.1. (i) $k_D(t,x|f_o(\cdot), f_o(\cdot),\dots,f_o(\cdot)) = 0$ if D is non-trivial (that is, not all ν_i vanish, cf. (5.1)).

(ii) The relation

$$\sum_{\rho+\sigma=\nu} \frac{1}{\rho!\sigma!}\, k_{D_1^\sigma \circ D_2^\rho}(t,x|f_o(\cdot),f_1(\cdot),f_o(\cdot)) = 0$$

holds identically in t,x, for $\nu = 1,2,\dots$. The summation has to be extended over all pairs of nonnegative integers ρ,σ such that $\rho+\sigma=\nu$. Note that for the function k_D appearing in this formula the number N is equal to 2, hence D can always be written as a formal product $D_1^\sigma \circ D_2^\rho$ (cf. (5.1)).

Proof. Statement (i) follows from part (ii) of Lemma 4.1. From parts (ii) and (iii) of the same lemma we obtain the identity (in t,x,ω)

$$c(t,x;\omega,\omega|f_o(\cdot),f_1(\cdot),f_o(\cdot)) = x .$$

The left hand side of this formula arises from

$c(t,x;w^1,w^2|f_o(\cdot),f_1(\cdot),f_o(\cdot))$ by means of the substitution $w^1 \to \omega$, $w^2 \to \omega$, so its derivatives with respect to ω can be expressed in terms of the derivatives of $c(\ldots)$ according to the Leibniz rule. If one differentiates the above relation ν-times with respect to ω and if ω is then replaced by t one arrives at the relation (ii).

The main objective of this section is to establish a recursive scheme, which allows to construct the k_D out of the given functions $f_o,\ldots,f_N$ by means of simple formal operations. This construction will be based on the following theorem

Theorem 5.1 (i) If $D = D_i^\nu$, $D_i = \partial/\partial w^i$, then $k_D(t,x) = h_i^{(\nu)}(t,x)$, where $h_i^{(\nu)}(t,x)$ is defined in Theorem 3.1. In this case k_D depends upon f_{i-1},f_i only.

(ii) Let $D = D' \circ D''$, where D'' does not act on $w^1,\ldots,w^M$ and D' does not act on $w^{M+1},\ldots,w^N$. Then

(5.3) $$k_D(t,x) = D'k_{D''}\Big(t,c(t,x;w^1,\ldots,w^M|f_o(\cdot),\ldots,f_M(\cdot))\Big|_{w\to w(t)}$$

where $k_*(t,x) = k_*(t,x|f_o(\cdot),\ldots,f_N(\cdot))$, $* = D,D''$.

Proof. Let $(\tilde{t},\tilde{x})$ be an arbitrary point of X and let $\tilde{x}(t)$ be the solution of $\dot{x} = f_o(t,x)$ satisfying $\tilde{x}(\tilde{t}) = \tilde{x}$. It follows then from (5.2) and from Theorem 4.1 that we have for an arbitrary D

(5.4) $(D\hat{x})(\tilde{w}) = k_D(\tilde{t},\tilde{x})$, where $\tilde{w} = w(\tilde{t})$.

If D is of the form D_i^ν then the expression on the left hand side of this relation is equal to $h_i^{(\nu)}(\tilde{t},\tilde{x})$, according to (3.3). This proves the first part of the theorem.

Assume now that D is factorized in the form $D' \circ D''$ where D',D'' have the properties stated in the theorem. k_D can then be calculated in the following way, according to the definition (5.2). First apply D'' on $c(\ldots)$ and put $w^{M+1}=\ldots=w^N=t$. This yields a function of $t,x,w^1,\ldots,w^M$. Let then D' act on this function and finally replace the remaining w^i by t. It follows now from Theorem 4.2 and from the definition (5.2) (with D" playing the role of D) that the result of the first procedure can be written as

$$k_{D''}(t,x'|f_M(\cdot),\dots,f_N(\cdot))$$

where x' is given by (4.6). In view of the second corollary of the same theorem this function is the same as

$$k_{D''}(t,x'|f_o(\cdot),\dots,f_N(\cdot)).$$

If we apply D' to what we just have obtained and if $w^1,\dots,w^M$ are then replaced by t we arrive at $k_D(t,x)$. This leads to the desired result (5.3).

We now consider a special factorization of

$$D = D' \circ D''$$

by assuming that D',D" satisfy these conditions

$$(5.5) \qquad \begin{array}{l} D' = D_i^\nu \\ D'' \text{ does not act on any } w^j \text{ for } j \le i \end{array}$$

(for the definition of D_i see (5.1)). If we use the first part of the theorem then one can describe the procedure leading to the expression on the right hand side of (5.3) also in these terms: Consider the composite function

$$(5.6) \qquad k_{D''}(t,c(\omega))$$

and regard $c(\omega)$ as an arbitrary, non specified function of the scalar variable ω. Differentiate (5.6) ν-times with respect to ω and represent the result as a function of the derivatives $c^{(\rho)}(\omega)$, $\rho = 0, \dots, \nu$. Replace then $c^{(\rho)}$ by $h_i^{(\rho)}(t,x)$. Note that this procedure makes sense also in case $\nu = 0$ and leads to $k_D = k_{D''}$, since $h_i^{(o)}t,x) = x$.

Comparing the above explanation with what was said in connection with (3.4) we are arrived at this result.

<u>Corollary.</u> Let D',D" satisfy condition (5.5), $\nu \ge 0$ being an integer, and let $D = D' \circ D''$. Then

(5.7) $k_D(t,x) = (\mathcal{D}_x^{\nu} k_{D''})(t,x_0,x_1,\ldots,x_\nu)\Big|_{x_\lambda \to h_i^{(\lambda)}(t,x),\lambda=0,\ldots,\nu}$.

In particular we have (cf. (3.5))

(5.8) $k_D(t,x) = (k_{D''})_x(t,x)\cdot h_i^{(1)}(t,x)$ in case $D = D_i \circ D''$ and D'' satisfies (5.5).

It is now clear how one can set up a recursive scheme in order to calculate k_D. Given D as a formal product (5.1), put

$$D^{(i)} = D_i^{\nu_i} \circ D_{i+1}^{\nu_{i+1}} \circ \cdots \circ D_N^{\nu_N}, \quad i = 1, \cdots, N.$$

We have then

(5.9) $D^{(i)} = D_i^{\nu_i} \circ D^{(i+1)}, \ i < N,$
$D' = D_i^{\nu_i},\ D'' = D^{(i+1)}$ satisfy condition (5.5).

Furthermore

$$D^{(N)} = D_N^{\nu_N}, \quad D^{(1)} = D .$$

It is now clear, in view of the corollary, how to compute recursively the functions $k_{D^{(N-1)}},\ldots,k_{D^{(1)}} = k_D$, starting with $k_{D^{(N)}} = h_N^{(\nu_N)}$.

As an application of the foregoing we state a lemma, which can be verified in a straightforward way using the corollary to Theorem 3.1.

Lemma 5.2. Let D be a differential operator of the form (5.1). Then the components of k_D can be written as polynomials in the partial derivatives of f_i for $i \le M$, where M is the largest integer i such that $\nu_i \neq 0$. The order of the derivatives which actually occur does not exceed

$$\sum_{j=1}^{N} \nu_j - 1 .$$

Furthermore for the calculation of k_D according to (5.2) N can be replaced by M (i.e. $k_D(t,x|f_0(\cdot),\ldots,f_N(\cdot)) = k_D(t,x|f_0(\cdot),\ldots,f_M(\cdot))$).

6. Linear Differential Equations.

In order to illustrate what we have done so far we will now consider in detail the case that all functions f_i are linear in x. So we assume throughout this section that $f_i(t,x)$ can be respresented in this form

$$(6.1) \qquad f_i(t,x) = F_i(t)x + b_i(t) ,$$

where F_i is a $n \times n$-matrix and b_i a n-dimensional column vector, $i = 0,\dots,N$. It will turn out that the calculation of the functions k_D becomes then much easier. This is mostly due to the fact that in the present situation differentiation of composite functions has only to be performed in case that the "outer function" is linear. Indeed, if $g(x) = Ax+l$ is a linear function in x, then the action of a differential operator D on a composite function $h(x)=g(f(x))$ is given by $(Dh)(x) = A\cdot(Df)(x)$. In particular it follows from (3.4) that $(\mathfrak{D}^{\nu}g)(x_o,\dots,x_\nu)=Ax_\nu, \nu > 0$. Hence the recursion formula in Theorem 3.1 assumes the form

$$(6.2) \qquad h_i^{(\nu+1)} = (h_i^{(\nu)})_t + (h_i^{(\nu)})_x\cdot(F_{i-1}x+b_{i-1})-F_i h_i^{(\nu)}$$

if $\nu \geq 1$. For $h_i^{(1)}$ we obtain from (3.10) the representation

$$(6.3) \qquad h_i^{(1)} = (F_{i-1}-F_i)x+b_{i-1}-b_i \quad .$$

It follows then from (6.2) and (6.3) that $h_i^{(\nu)}$ is also linear in x and hence can be written in the form

$$(6.4) \qquad h_i^{(\nu)} = G_i^{(\nu)}x+l_i^{(\nu)} \quad .$$

$G_i^{(\nu)} = G_i^{(\nu)}(t)$ and $l_i^{(\nu)} = l_i^{(\nu)}(t)$ respectively are time-dependent matrices and column vectors respectively. From (6.2) one can easily derive a system of recursion formulas for $G_i^{(\nu)}$ and $l_i^{(\nu)}$. It is also not difficult to obtain from Definition 4.1 the following general result.

(6.5) **If (6.1) is true then $c(t,x;w|f_o(\cdot),\dots,f_N(\cdot))$ is a linear function of the state variable x.**

This follows simply from the fact that $x_i(t;t_o,x)$ and hence also $\hat{y}_i(\cdot)$ (cf. (4.1)) is a linear function of x. Using the definition (5.2) of k_D and Theorem 5.1 the statement of the following theorem is then immediately verified, in view of (6.2)-(6.5).

<u>Theorem 6.1.</u> If f_i is of the form (6.1) for $i=0,\ldots,N$ then all functions k_D are linear in x and hence can be written in the form

$$k_D(t,x) = K_D(t)x + \tilde{k}_D(t)$$

where $K_D, \tilde{k}_D$ respectively are t-dependent matrices and column vectors respectively. In case $D = D_i^\nu$ we write $G_i^{(\nu)}$ and $1_i^{(\nu)}$ respectively instead of $K_D, \tilde{k}_D$ respectively (in accordance with our previous custom). K_D and $\tilde{k}_D$ can be computed from the given quantities F_i, b_i (cf. (6.1)) by means of the subsequent set of recursive relations (the argument t has been omitted everywhere)

(i) $$G_i^{(1)} = F_{i-1} - F_i, \quad 1_i^{(1)} = b_{i-1} - b_i,$$

$$G_i^{(\nu+1)} = \dot{G}_i^{(\nu)} + G_i^{(\nu)} F_{i-1} - F_i G_i^{(\nu)} \quad ,$$

$$1_i^{(\nu+1)} = \dot{1}_i^{(\nu)} + G_i^{(\nu)} b_{i-1} - F_i 1_i^{(\nu)} \quad .$$

(ii) If the differential operator D is factorized in the form $D' \circ D''$ where D', D'' satisfy condition (ii) of Theorem 5.1 then

$$K_D = K_{D''} \cdot K_{D'} \quad , \quad \tilde{k}_D = K_{D''} \tilde{k}_{D'} \quad .$$

7. Control Systems.

We now turn to the study of control systems, which is the subject we are mostly interested in this work. By a control system we mean a differential equation

(7.1) $\dot{x} = f(t,x;u)$, $f = (f^1,f^2,\ldots,f^n)^T$

where the function f is defined and of class C^∞ on an open set Y in (t,x,u)-space. The n-dimensional vector $x = (x^1,\ldots,x^n)^T$ is called the state variable, the m-dimensional vector $u = (u^1,\ldots,u^m)^T$ is called the control variable.

Let X be an open, non-empty set in (t,x)-space and I an open t-interval. Assume that we have N+1 functions $u_i(t)$ which satisfy the following conditions for $i=0,\ldots,N$.

$u_i(t)$ is an m-dimensional vector, the components of $u_i(\cdot)$
(7.2) are C^∞ functions of t on I,
$(t,x)\in X$ implies $t\in I$ and $(t,x,u_i(t))\in Y$.

If we put

(7.3) $$f_i(t,x) = f(t,x;u_i(t)) ,$$

$i=0,\ldots,N$, we obtain N+1 functions of (t,x) which are defined and of class C^∞ on X. The function $f(t,x|w)$, which is associated with the system (7.3) according to the definition (2.3) can then always be written in the form $\dot{x} = f(t,x;u_w(t))$, where $u_w(t)$ is - for fixed w - a piecewise C^∞-function of t. For the time being such a function will be called an admissible control function; later we will consider situations where "admissible" includes constraints of the form $u(t)\in U$. By an admissible trajectory we always mean a solution of a differential equation $\dot{x} = f(t,x;u(t))$, where u(t) is an admissible control function.

Assuming that the f_i are of the form (7.3) the values of the function $\hat{x}(w) = \hat{x}(w;\tilde{x}(t_o))$ represent terminal points of admissible trajectories if $w \in W$ and $\|w-\tilde{w}\|$ is sufficiently small. To be more specific, they belong to the set $\mathcal{A}$ of all points attainable at $t=\tilde{t}$ along admissible trajectories initiating at time t_o from $\tilde{x}(t_o)$. Here $\tilde{x}(t)$ is regarded as a fixed solution of the eq. $\dot{x}=f(t,x;u_o(t))$ and will be denoted henceforth as reference trajectory; $u_o(\cdot)$ will be called reference control. It is clear that we obtain points of $\mathcal{A}$ in the neighborhood of $\tilde{x}(\tilde{t})$ not only by varying the parameter w, but also by considering all possible choices of N and of the control functions $u_1,\ldots,u_N$. The behaviour of $\hat{x}(w)$ if all these data are variable is important for the study of the local properties of $\mathcal{A}$ which is the main objective of this paper. Therefore the present section is devoted to a review of the previous material, taking now into account that the functions f_i arise from a "universal" function $f(t,x;u)$ by means of a substitution $u \to u_i(t)$. It will turn out that a similar statement holds with respect to the functions k_D: There exists a "universal"

$$k_D(t,x;\dot{u}_1,\dot{u}_1,\ddot{u}_1,\ldots,u_2,\dot{u}_2,\ddot{u}_2,\ldots,u_3,\dot{u}_3,\ddot{u}_3,\ldots\ldots\ldots\ldots)$$

from which the particular k_D can be obtained by inserting $u_i(t)$, $\dot{u}_i(t)$, $\ddot{u}_i(t)$ ect. for $u_i,\dot{u}_i,\ddot{u}_i$ ect. respectively. We now wish to make this statement more precise and establish some basic relations for this "universal" functions. For this purpose it is convenient to introduce a special notation. By the symbol $\boldsymbol{u}$ we denote from now on a sequence of m-dimensional vectors u_i:

$$(7.4) \qquad \boldsymbol{u} = \{u_o,u_1,u_2,\ldots,u_k,\ldots\ \ldots\} .$$

The components of each u_i will be denoted by $u_i^{(\mu)}$. Whether the $u_i^{(\mu)}$ have to be regarded as independent variables or as given real numbers will become clear from the context. If $u(\cdot)$ is an admissible control function and if t is a real number we denote by $\boldsymbol{u}(t)$ the sequence

(7.5) $\{u(t),\dot{u}(t),\ddot{u}(t),\ldots, d^k u(t)/dt^k,\ldots\}$.

Occasionally it is convenient to have a given $\mathbf{U}$ represented in the form (7.5). We will then use the following lemma which can be verified immediately.

<u>Lemma 7.1.</u> Let $\mathbf{U}$ be a sequence of the form (7.4) and let all but finitely many u_ν be equal to zero. Then

(7.6) $$u(t,\tau;\mathbf{U}) := \sum_{\nu=0}^{\infty} \frac{u_\nu}{\nu!}(t-\tau)^\nu$$

is well defined for all t,τ and is an admissible control function (as a function of t).Furthermore the following statements hold (i) the sequence

(7.7) $$\hat{\mathbf{U}}(t,\tau;\mathbf{U}) := \{u(t,\tau;\mathbf{U}), \frac{\partial}{\partial t}u(t,\tau;\mathbf{U}), \frac{\partial^2}{\partial t^2}u(t,\tau;\mathbf{U}),\ldots\}$$

Coincides with $\mathbf{U}$ whenever $t = \tau$.

(ii) If $(\tilde{t},\tilde{x},u_o)\in Y$ then there exists an $\epsilon>0$ and a neighborhood $\mathcal{N}$ of $(\tilde{t},\tilde{x})$ (in the (t,x)-space) such that

$$(t,x,u(t,\tau;\mathbf{U}))\in Y$$

whenever $(t,x)\in\mathcal{N}$ and $|t-\tau|\le\epsilon$.

We are now in a position to give a precise meaning to the notion of the "universal" k_D. These functions will be denoted by $K_D(t,x,\mathbf{U}_o,\mathbf{U}_1,\ldots)$ and are n-dimensional vectors depending upon t,x and finitely many out of an infinite set of further variables, which we denote by $u_{i,\rho}$ and which are m-dimensional vectors. For notational convenience we arrange these variables in sequences

(7.8) $\mathbf{U}_i = \{u_{i,o}, u_{i,1},\ldots, u_{i,k},\ldots\}$, $i=0,1,\ldots$.

The usage of the symbol $\mathbf{U}_i$ is in accordance with the notation introduced earlier (cf. (7.4)). It will turn out that K_D depends upon $\mathbf{U}_o,\ldots,\mathbf{U}_N$ only if D admits a factorization of the form

(5.1) We therefore denote this function then by $K_D(t,x,\mathbf{u}_o,\mathbf{u}_1\ldots,\mathbf{u}_N)$

The explicit definition of K_D will now be given pointwise. We specify the value of this function for any particular point $(\tilde{t},\tilde{x},\mathbf{u}_o,\mathbf{u}_1,\ldots.)$ of the set

(7.9) $\{t,x,\mathbf{u}_o,\mathbf{u}_1,\ldots. : (t,x,u_{i,o})\in Y \text{ for } i=0,1,\ldots\}$

in the following way.

First let us assume that all but finitely many components of $\mathbf{u}_i$ vanish, for each $i=0,1,\ldots.$ As can be seen from Lemma 7.1 the function

(7.10) $\tilde{f}_i(t,x) = f(t,x;u(t,\tilde{t};\mathbf{u}_i))$

is defined and of class C^∞ on a neighborhood of $(\tilde{t},\tilde{x})$, for $i=0,1,\ldots$ We then choose a factorization (5.1) of D and associate with the $\tilde{f}_i$ the function $k_D(t,x|\tilde{f}_o(\cdot),\ldots,\tilde{f}_N(\cdot))$ according to the construction described in Section 5. Finally we put

(7.11) $K_D(\tilde{t},\tilde{x};\mathbf{u}_o,\mathbf{u}_1,\ldots) = k_D(\tilde{t},\tilde{x}|\tilde{f}_o(\cdot),\ldots,\tilde{f}_N(\cdot))$.

The extension of the definition to an arbitrary point $(t,x,\mathbf{u}_o,\ldots)$ of the set (7.9) will be based on the following lemma.

<u>Lemma 7.2.</u> Let D be a differential operator of the form (5.1) and let $M = M(D)$ be the largest integer i such that $\nu_i \neq 0$. Furthermore let

(7.12) $$\nu(D) = \sum_{j=1}^{N} \nu_j .$$

Then $K_D(t,x;\mathbf{u}_o,\mathbf{u}_1,\ldots)$ actually depends upon t,x and those $u_{i,\rho}$ only for which

(7.13) $i \leq M(D), \quad \rho \leq \nu(D)-1$.

K_D is a C^∞ function of t,x and the components of $\mathbf{u}_o,\ldots,\mathbf{u}_M$. It can be represented as a polynomial in the components of $\mathbf{u}_1,\ldots\mathbf{u}_M$

<u>Proof.</u> If the function $f(t,x;u(t,\tau;\mathbf{u}))$ is differentiated σ-times with respect to t and the components of x and if t is then replaced by τ, one obtains a polynomial in $u_1,\ldots,u_\sigma$, the coefficients being partial derivatives of $f(t,x;u)$ at $(t,x,u) = (\tau,x,u_o)$. This follows immediately from (7.6). Combining this statement with the definition (7.10) and (7.11) of K_D the desired result becomes an

obvious consequence of Lemma 5.2.

We now are in a position to explain the meaning of $K_D(t,x;\,\mathbf{U}_o,\mathbf{U}_1,\dots)$ for an arbitrary element of the set (7.9). We simply define

$$K_D(t,x;\mathbf{U}_o,\mathbf{U}_1,\dots) := K_D(t,x;\,\mathbf{U}'_o\,,\;\mathbf{U}'_1\,,\;\dots\dots), \tag{7.14}$$

where $\mathbf{U}'_i$ arises from $\mathbf{U}_i = \{u_{i,o},\dots\dots\}$ by putting $u_{i,\rho} = 0$ if $i > M(D)$ or $\rho > \nu(D)-1$.

The functions K_D will play the key role in the later parts of this paper. Before we turn to the study of their properties we wish to describe more precisely the relation of K_D to the functions k_D introduced in Section 5.

<u>Lemma 7.3</u> Let $u_i(\cdot)$ satisfy condition (7.2) and let f_i be defined according to (7.3), for $i=0,\dots,N$. Then the identity

$$\begin{aligned} &k_D(t,x\,|\,f_o(\cdot),f_1(\cdot),\dots,f_N(\cdot)) = \\ &\qquad = K_D(t,x;\,\mathbf{U}_o(t),\,\mathbf{U}_1(t),\dots,\,\mathbf{U}_N(t)) \end{aligned} \tag{7.15}$$

holds on X for every D of the form (5.1). Here $\mathbf{U}_i(t)$ denotes the sequence of the form (7.5) which is associated with the function $u(\cdot) = u_i(\cdot)$.

<u>Proof</u> Let D be a fixed differential operator of the form (5.1) and let $(\tilde{t},\tilde{x})$ be a fixed point of X. In order to calculate $k_D(\tilde{t},\tilde{x})$ we can replace each $u_i(t)$ by its Taylor polynomial of degree $\nu(D)-1$ at $\tilde{t}$, in view of Lemma 5.2. Also, because of Lemma 7.2, all terms of the sequence $\mathbf{U}_i(\tilde{t})$ which carry a subscript bigger than $\nu(D)-1$ do not enter into the explicit formula for $K_D(\tilde{t},\tilde{x};\;\mathbf{U}_o(\tilde{t}),\dots)$. In other words, for the purpose of the proof we may assume that each $u_i(\cdot)$ is a polynomial in t. It follows then from (7.5) and (7.6) that the relation

$$u_i(t,\tau;\;\mathbf{U}_i(\tau)) = u_i(t)$$

holds identically in t,τ and for $i=0,\dots,N$. Hence the functions f_i and $\tilde{f}_i$ coincide, if $\mathbf{U}_i = \mathbf{U}_i(\tilde{t})$ (cf. (7.3) and (7.10)). For $t = \tilde{t}, x=\tilde{x}$ (7.15) therefore becomes an immediate consequence of (7.11).

Next we wish to describe an algorithm for the construction of the functions K_D which formally is more easy to handle than the definition. To this purpose let

$$\mathcal{U} = \{u_o, u_1, \dots\} \quad \text{and} \quad \mathcal{V} = \{v_o, v_1, \dots\}$$

be two sequences of independent variables u_i, v_i, each variable being a m-dimensional vector. We now introduce functions $h^{(\nu)}(t,x;\mathcal{U},\mathcal{V})$, $\nu = 0,1,\dots$ Each $h^{(\nu)}$ is a n-dimensional vector, depends upon t,x and finitely many components of $\mathcal{U}$ and $\mathcal{V}$ and is a C^∞ function of all its arguments on the set

$$\{t,x,\mathcal{U},\mathcal{V} \; : \; (t,x,u_o) \in Y \text{ and } (t,x,v_o) \in Y\} .$$

The definition is recursively and goes as follows

$$h^{(o)}(t,x;\mathcal{U},\mathcal{V}) = x,$$

$$\begin{aligned} h^{(\nu+1)}(t,x;\mathcal{U},\mathcal{V}) = (h^{(\nu)})_t &+ \sum_{\sigma=0}^{\infty} (h^{\nu})_{u_\sigma} \cdot u_{\sigma+1} + \\ &+ \sum_{\sigma=0}^{\infty} (h^{(\nu)})_{v_\sigma} \cdot v_{\sigma+1} + (h^{(\nu)})_x \cdot f(t,x;u_o) - \\ &- (\mathcal{D}_x^{\nu} f)(t,x_o,\dots,x_\nu;v_o)\Big|_{x_\lambda \to h^{(\lambda)}} \end{aligned} \tag{7.16}$$

Here $(h^{(\nu)})_x, (h^{(\nu)})_{u_\sigma}$ etc. are the Jacobian matrices of $h^{(\nu)}$ with respect to x, u_σ etc. The argument in the elements of these matrices as well as in $h^{(\lambda)}$ is of course $t,x,\mathcal{U},\mathcal{V}$. The last term in the recursion formula arises from the function $f(t,x;v_o)$ by applying the operator $\mathcal{D}_x^{\nu}$ (cf. (3.4)) and then replacing the variable x_λ by $h^{(\lambda)}$ for $\lambda = 0,\dots,\nu$. It can be seen from (7.16) that $h^{(\nu)}$ depends upon u_σ, v_σ for $\sigma < \nu$ only, hence the infinite sums are in fact finite.

We have in particular

$$h^{(1)}(t,x;\mathbf{U},\mathbf{V}) = f(t,x;u_o)-f(t,x;v_o),$$

$$(7.17)\quad h^{(2)}(t,x;\mathbf{U},\mathbf{V}) = (h^{(1)})_t+f_u(t,x;u_o)\cdot u_1-f_u(t,x;v_o)\cdot v_1 + (h^{(1)})_x\cdot f(t,x;u_o)-f_x(t,x;v_o)\cdot h^{(1)}.$$

We now state the main result of this section

Theorem 7.1 (i) If $D = D_i^\nu$, $D_i = \partial/\partial w^i$, then $K_D(t,x;\mathbf{U}_o,\mathbf{U}_1,\ldots) = h^{(\nu)}(t,x;\mathbf{U}_{i-1},\mathbf{U}_i)$.

(ii) If $D = D'\circ D''$ and if D', D'' satisfy (5.5), then $K_D(t,x;\mathbf{U}_o,\mathbf{U}_1,\ldots)$ is equal to

$$(\mathcal{D}_x^\nu K_{D''})(t,x_o,x_1,\ldots,x_\nu;\mathbf{U}_o,\mathbf{U}_1,\ldots.)\Big|_{\substack{x_\lambda\to h^{(\lambda)}(t,x;\mathbf{U}_{i-1},\mathbf{U}_i),\\ \lambda=0,\ldots,\nu}}$$

Proof. It is sufficient to prove the theorem under the additional assumption $\nu(D) \le L$, where L is a fixed but arbitrary positive integer (for the definition of $\nu(D)$ see (7.12)). It is then clear, in view of Lemma 7.2, that we can assume for the purpose of the proof that in each sequence $\mathbf{U}_j$ all members $u_{j,\sigma}$ are equal to zero, if $\sigma > L$. The remaining members have to be regarded as fixed vectors for the rest of the proof. It then turns out that the functions $u(t,\tau;\mathbf{U}_j)$ are polynomials in t and τ. It follows therefore from Lemma 7.3 that the identity

$$(7.18)\quad \begin{aligned} &K_D(t,x;\hat{\mathbf{U}}(t,\tau;\mathbf{U}_o),\hat{\mathbf{U}}(t,\tau;\mathbf{U}_1),\ldots\ldots) = \\ &= k_D(t,x|f_o(\cdot,\tau),f_1(\cdot,\tau),\ldots,f_N(\cdot,\tau)) \end{aligned}$$

holds on the set

$$\{t,\tau,x : (t,x,u(t,\tau;\mathbf{U}_j)) \in Y \text{ for } j=0,\ldots,N\}.$$

Here we denote by $f_j(\cdot,\tau)$ the function given by

$$f_j(t,x,\tau) := f(t,x;u(t,\tau;\mathbf{U}_j)).$$

For the definition of $u(t,\tau;\mathbf{U})$, $\hat{\mathbf{U}}(t,\tau;\mathbf{U})$ respectively see (7.6) and (7.7) respectively.

Part (i) : $D = D_i^\nu$.

We regard i as fixed and denote for the moment the left hand side of (7.18) - with $D = D_i^\nu$ - by $K_\nu(t,\tau,x)$. Using the representation of K_ν given by the right hand side of (7.18) one can infer from Theorem 5.1 and Theorem 3.1 the following system of recursive relations

$$(7.19)\qquad \begin{aligned} K_o(t,\tau.x) &= x \\ K_{\nu+1}(t,\tau,x) &= \frac{\partial}{\partial t} K_\nu + (K_\nu)_x \cdot f(t,x;u(t,\tau;\mathbf{U}_{i-1})) \\ &\quad - (\mathcal{D}_x^\nu f)(t,x_o,\ldots,x_\nu;u(t,\tau;\ \mathbf{U}_i))\Big|_{x_\lambda \to K_\lambda(t,\tau,x)}. \end{aligned}$$

We now proceed in proving the first part of the theorem by induction with respect to ν. So we may assume that the assertion has been established for a certain $\nu < L$. This implies in particular that we have

$$K_\lambda(t,\tau,x) = h^{(\lambda)}(t,x;\ \hat{\mathbf{U}}(t,\tau;\ \mathbf{U}_{i-1}),\ \hat{\mathbf{U}}(t,\tau;\ \mathbf{U}_i))$$

for all $\lambda \le \nu$. Taking the special form (7.7) of $\hat{\mathbf{U}}(t,\tau;\mathbf{U})$ into account we obtain from the last relation (for $\nu = \lambda$)

$$\frac{\partial}{\partial t} K_\nu(t,\tau,x) = (h^{(\nu)})_t + \sum_{\sigma=0}^{\infty} (h^{(\nu)})_{u_\sigma} \cdot \frac{\partial^{\sigma+1}}{\partial t^{\sigma+1}} u(t,\tau;\mathbf{U}_{i-1})$$

$$\sum_{\sigma=0}^{\infty} (h^{(\nu)})_{v_\sigma} \cdot \frac{\partial^{\sigma+1}}{\partial t^{\sigma+1}} u(t,\tau;\ \mathbf{U}_i)\ .$$

The argument in $(h^{(\nu)})_t$ etc. is of course

$$t,x,\ \hat{\mathbf{U}}(t,\tau;\ \mathbf{U}_{i-1}),\ \hat{\mathbf{U}}(t,\tau;\ \mathbf{U}_i)\ .$$

The proof by induction is easily completed with the help of the recursive relation (7.16) and of Lemma 7.1, part (i). Indeed, if we put $\tau = t$ and substitute what we then obtain from the above formulas for K_λ and $\partial K_\nu/\partial t$ into the right hand side of (7.19) we get

$$K_{\nu+1}(t,t,x) = h^{(\nu+1)}(t,x;\ \mathbf{U}_{i-1},\ \mathbf{U}_i).$$

On the other hand we have $\hat{\mathbf{U}}(t,t;\mathbf{U}) = \mathbf{U}$ and therefore the expression on the left hand side of the last relation is nothing else than

$K_D\,(t,x;\ \mathcal{U}_o,\ \mathcal{U}_1,\ldots.)$ for $D = D_i^{\nu+1}$.

Part (ii) : $D = D' \circ D''$, where D', D'' satisfy (5.5).

We apply the corollary to Theorem 5.1, with $f_i(\cdot,\tau)$ playing the role of $f_i(\cdot)$. Using (7.18) and the first part of Theorem 7.1 one sees that in the present situation the relation (5.7) can be written in this form

$$K_D(t,x;\ \hat{\mathcal{U}}(t,\tau;\ \mathcal{U}_o),\ \hat{\mathcal{U}}(t,\tau;\ \mathcal{U}_1),\ldots\ldots) =$$
$$= (\mathcal{D}\,{}_x^{\nu}K_{D''})(t,x_o,\ldots,x_\nu;\ \hat{\mathcal{U}}(t,\tau;\ \mathcal{U}_o),\ \hat{\mathcal{U}}(t,\tau;\ \mathcal{U}_1),\ldots)$$

with $x_\lambda \to h^{(\lambda)} = h^{(\lambda)}(t,x;\ \hat{\mathcal{U}}(t,\tau;\ \mathcal{U}_{i-1}),\ \hat{\mathcal{U}}(t,\tau;\ \mathcal{U}_i))$.

This relation can be given the form stated in the theorem simply by putting $\tau = t$ and using the identity $\hat{\mathcal{U}}(t,t;\ \mathcal{U}) = \mathcal{U}$ once more.

Comparing Theorem 5.1 and Theorem 7.1 one sees immediately that the method for computing k_D, which was outlined at the end of Sec. 5, verbally carries over to the functions K_D.

We conclude this section with a further lemma, which will be needed subsequently.

Lemma 7.4 The following relations hold identically in $t,x,\ \mathcal{U}_o, \mathcal{U}_1$

(i) $K_D(t,x;\ \mathcal{U}_o,\ \mathcal{U}_1,\ldots.) = 0$ if $\mathcal{U}_i = \mathcal{U}_o$ for $i=1,2,\ldots$ and if $\nu(D) > 0$.

(ii) $\displaystyle\sum_{\sigma+\rho=\nu} \frac{1}{\sigma!\rho!} K_{D_1^\sigma \circ D_2^\rho}(t,x;\ \mathcal{U}_o,\ \mathcal{U}_1,\ \mathcal{U}_o) = 0$ if $\nu > 0$.

The proof follows immediately from Lemma 5.1 and from our definition of K_D (see (7.10), (7.11)).

8. Formal Taylor Expansions. The General Composition Rule.

We continue the discussion of control systems defined by an equation of the form (7.1) and use all the notation introduced so far. In contrast to previous custom the positive integer N has to be regarded as fixed throughout this section. The symbol D will then always refer to a product of formal powers $D_i^{\nu_i}$, $i=1,\dots,N$, and therefore can completely be described by the N-tuple $(\nu_1,\dots,\nu_N)$. For this reason we write

$$K_{\underline{\nu}}(t,x;\; \mathcal{U}_0,\dots,\; \mathcal{U}_N) \quad\text{or}\quad K_{\nu_1,\dots,\nu_N}(t,x;\; \mathcal{U}_0,\dots,\; \mathcal{U}_N)$$

instead of K_D if D admits the representation (5.1).

Let $z_1,\dots,z_N$ be N scalar variables. We introduce the following abbreviations

$$z \text{ for } (z_1,\dots,z_N)^T, \qquad \underline{\nu} \text{ for } (\nu_1,\dots,\nu_N)^T$$

$$z^{\underline{\nu}} \text{ for } \prod_{i=1}^{N} z_i^{\nu_i}, \qquad |\underline{\nu}| \text{ for } \sum_{i=1}^{N} \nu_i, \quad \underline{\nu}! \text{ for } \prod_{i=1}^{n} (\nu_i!), \tag{8.1}$$

and $C(t,x,z_1,\dots,z_N;\; \mathcal{U}_0,\; \mathcal{U}_1,\dots,\; \mathcal{U}_N)$ or $C(t,x,z;\; \mathcal{U}_0,\dots,\; \mathcal{U}_N)$ for the formal power series

$$x + \sum_{|\underline{\nu}|>0} \frac{1}{\underline{\nu}!} K_{\underline{\nu}}(t,x;\; \mathcal{U}_0,\dots,\; \mathcal{U}_N) z^{\underline{\nu}} \,. \tag{8.2}$$

The infinite sum is formal and has to be extended over all $\underline{\nu}$ with nonnegative integer components ν_i which do not vanish simultaneously. Note that C is actually a n-dimensional column vector of formal power series, since the $K_{\underline{\nu}}$ are such column vectors and $z^{\underline{\nu}}$ is scalar.

Intuitively it is clear that C is related to the Taylor expansion of some functions which arise as solutions of our control problem, in view of the relation between K_D and k_D. For our purposes we have to make this connection more precise. This is done in

<u>Lemma 8.1</u> Let $u_0(\cdot),\dots,u_N(\cdot)$ be C^∞ functions of t and assume that the condition

(8.3) $\quad (\tilde{t},\tilde{x},u_i(\tilde{t}))\in Y \quad$ for $\ i=0,\ldots,N$

is satisfied for some $\tilde{t},\tilde{x}$. Let the function $f_i(t,x)$ and the sequence $\mathcal{U}_i(t)$ respectively be associated with $u_i(\cdot)$ according to (7.3) and (7.5) respectively for $i=0,\ldots,N$. Finally, let $c(t,x;w|f_o(\cdot),\ldots,f_N(\cdot))$ have the meaning explained in Definition 4.1, where $f_i(\cdot)$ stands for the function just introduced. Then the power series $C(\tilde{t},\tilde{x},z_1,\ldots,z_N;\ \mathcal{U}_o(\tilde{t}),\ldots,\mathcal{U}_N(\tilde{t}))$ is the Taylor-expansion at $z=0$ of the function

$$c(\tilde{t},\tilde{x};w(\tilde{t})+z|f_o(\cdot),\ldots,f_N(\cdot)) =$$

(8.4) $$= c(\tilde{t},\tilde{x};\tilde{t}+z_1,\ldots,\ \tilde{t}+z_N|f_o(\cdot),\ldots,f_N(\cdot)).$$

Proof. It is clear, in view of (8.3), that all $f_i(\cdot)$ are well defined and of class C^∞ in some neighborhood of $(\tilde{t},\tilde{x})$. From the results of Sec. 4 it follows then that the expression appearing in (8.4) represents a well defined and infinitely often differentiable function of z in a certain neighborhood of $z=0$. Furthermore, if a differential operator D which acts on z is applied to this expression and if z is then put equal to zero one obtains

$$k_D(\tilde{t},\tilde{x}|f_o(\cdot),\ldots,f_N(\cdot)) = K_D(\tilde{t},\tilde{x};\ \mathcal{U}_o(\tilde{t}),\ldots,\ \mathcal{U}_N(\tilde{t}))$$

(cf. (5.2) and (7.15)). Taking into account the notational changes indicated at the beginning of this section the statement of the lemma becomes an immediate consequence of the definition (8.2).

Before we continue with the discussion of the formal properties of the power series C we wish to point out that the n-dimensional column vector c defined by (8.4) can be interpreted as terminal point of an admissible trajectory whenever the components of z satisfy the inequalities

(8.5) $\quad z_1 \leq z_2 \leq \ldots \leq z_N \leq 0$.

Indeed, if (8.5) holds the following definition makes sense and yields - for fixed z - an admissible control function

$$(8.6)\quad u_z(t) = \begin{cases} u_o(t) & \text{if } t \le \tilde{t}+z_1 , \\ u_i(t) & \text{if } \tilde{t}+z_i < t \le \tilde{t}+z_{i+1},\ i=1,\dots,N-1, \\ u_N(t) & \text{if } t > \tilde{t}+z_N. \end{cases}$$

Consider now the solution $\tilde{x}(t)$ of the initial value problem

$$\dot{x} = f(t,x;u_o(t)), \quad x(\tilde{t}) = \tilde{x},$$

which exists on some interval $[t_o,\tilde{t}]$. One sees then immediately from Theorem 4.1 and from our considerations in Sec. 2 that the following statement holds true. If z satisfies (8.5) and if $\|z\|$ is sufficiently small then the solution of the initial value problem

$$(8.7)\qquad \dot{x} = f(t,x;\ u_z(t)), \quad x(t_o) = \tilde{x}(t_o)$$

exists on $[t_o,\tilde{t}]$ and assumes the value (8.4) at $t = \tilde{t}$.

Next the main tool for the evaluation of the higher order necessary conditions will be presented. We call it the general composition rule since it describes a certain scheme by which a power series of the form (8.2) can be composed out of power series of the same type but with a lower number of variables.

<u>Theorem 8.1 (General Composition Rule).</u> The following identity holds for every integer M such that $1 \le M < N$:

$$(8.8)\qquad \begin{aligned} &C(t,x,z_1,\dots,z_N;\ \mathcal{U}_o, \mathcal{U}_1,\dots, \mathcal{U}_N) = \\ &= C(t,x',z_{M+1},\dots,z_N;\ \mathcal{U}_M, \mathcal{U}_{M+1},\dots, \mathcal{U}_N) \\ &\text{with } x' = C(t,x,z_1,\dots,z_M;\ \mathcal{U}_o, \mathcal{U}_1,\dots, \mathcal{U}_M) \ . \end{aligned}$$

<u>Remark</u>. A careful interpretation of the procedure leading to the expressions in this formula seems to be in order. This procedure actually involves two formal power series, namely

$$(8.9)\qquad \begin{aligned} &C(t,x+y,\ z_{M+1},\dots,z_N;\ \mathcal{U}_M,\dots, \mathcal{U}_N) = x+y\ + \\ &+ \sum\varphi(\nu_{M+1},\dots,\nu_N)K_{\nu_{M+1},\dots,\nu_N}(t,x+y;\ \mathcal{U}_M,\dots, \mathcal{U}_N)z_{M+1}^{\nu_{M+1}}\dots z_N^{\nu_N} \end{aligned}$$

and $C(t,x,z_1,\dots,z_M;\ \mathcal{U}_o,\dots,\mathcal{U}_M) = x+y$, where

$$(8.10)\quad y = \sum \varphi(\nu_1,\dots,\nu_M)K_{\nu_1,\dots,\nu_M}(t,x;\ \mathcal{U}_o,\dots, \mathcal{U}_M)z_1^{\nu_1}\dots z_M^{\nu_M} \ .$$

The sum in (8.9) has to be extended over all $\nu_{M+1},\dots,\nu_N$ such that $\nu_i \geq 0$ and $\sum \nu_i > 0$, the sum in (8.10) over all $\nu_1,\dots,\nu_M$ such that $\nu_i \geq 0$ and $\sum \nu_i > 0$. Furthermore we have used the abbreviation $\varphi(\nu_i,\dots,\nu_j) = (\nu_i!)\cdots(\nu_j!)$.

Assume now that $t, x, \mathbf{u}_0,\dots,\mathbf{u}_N$ are fixed and belong to the set (7.9). It follows then from the considerations of the previous section that each $K_{\dots}(t,x+y;\mathbf{u}_M,\dots,\mathbf{u}_N)$, regarded as a function of y, is of class C^∞ in a neighborhood of $y=0$ and hence admits a Taylor-expansion which leads to a (formal) power series with respect to y. If y is then replaced by the power series (8.10) each $K_{\dots}(t,x+y;\mathbf{u}_M,\dots,\mathbf{u}_N)$ turns into a power series in $z_1,\dots,z_M$. Finally, substituting for each $K_{\dots}$ the corresponding power series into the second line of (8.9) one arrives at a double series which has then to be rearranged in such a way that the outcome will be a power series in $z_1,\dots,z_N$: This is what we actually mean by the second and third line of the formula (8.8). It is not difficult to see that in carrying out this procedure one never will encounter formal difficulties or face questions of convergence. Indeed, if one picks a certain $\underline{\nu}$ and examines the coefficient of $z^{\underline{\nu}}$ in the power series obtained at the end one sees the following. The components of this coefficient can be written as polynomials in quantities which themselves can be derived by means of rational operations and differentiation with respect to x from those coefficients of (8.9) and (8.10) respectively for which one has $\nu_{M+1}+\dots+\nu_N \leq |\underline{\nu}|$ and $\nu_1+\dots+\nu_M \leq |\underline{\nu}|$.

<u>Proof of Theorem 8.1.</u> First we wish to convince ourselves that it is sufficient to prove the theorem under the additional hypothesis that all but finitely many terms $u_{i,\rho}$ in each sequence (7.8) which occurs in the statement of the theorem are zero. Indeed, given a positive integer $\varkappa$, the coefficients of $z^{\underline{\nu}}$ for $|\underline{\nu}| \leq \varkappa$ in any power series of the form (8.2) do not depend upon $u_{i,\rho}$ for $\rho \geq \varkappa$

according to Lemma 7.2. From this statement and from the above remark one arrives immediately at this conclusion: The question whether the coefficients of z^{ν} in the two power series appearing in (8.8) are the same or not remains unaffected from the choice of the $u_{i,\rho}$ for $\rho \geq \varkappa$.

Hence we assume from now on that all but finitely many components of each $\mathfrak{U}_i$ are zero. The polynomials $u(t,\tau;\ \mathfrak{U}_i)$ are then well defined for all t,τ (cf. (7.6.)). We regard t,x and the $\mathfrak{U}_i$ as fixed throughout the remaining part of the proof and write $\tilde{t},\tilde{x}$ instead of t,x in order to avoid notational confusion.

Let us now consider the functions $\tilde{f}_i(t,x)$ which were introduced by (7.10). They are well defined and of class C^∞ in a neighborhood of $(\tilde{t},\tilde{x})$. It follows therefore from Theorem 4.2 that for $t=\tilde{t},x=\tilde{x}$, $f_i = \tilde{f}_i$, $i = 0,\dots,N$, the relations (4.5), (4.6) hold identically in w in a neighborhood of $w=w(\tilde{t})$ and that the expressions occuring in these relations are C^∞-functions of w. On the other hand we know from Lemma 8.1 that the Taylor-expansion at $z=0$ of the function

$$c(\tilde{t},x;w(\tilde{t})+z\,|\,\tilde{f}_0(\cdot),\dots,\tilde{f}_N(\cdot))$$

is given by the formal power series

$$C(\tilde{t},x,z;\ \mathfrak{U}_0(\tilde{t}),\ \mathfrak{U}_1(\tilde{t}),\dots,\ \mathfrak{U}_N(\tilde{t}))$$

where

$$\mathfrak{U}_i(\tilde{t}) = \{u(\tilde{t},\tilde{t};\mathfrak{U}_i),\ \frac{\partial u}{\partial t}(\tilde{t},\tilde{t};\ \mathfrak{U}_i),\dots\dots\} = \mathfrak{U}_i$$

(cf. (7.7)). Hence the statement of the theorem is a consequence of Theorem 4.2 and can be obtained from (4.5), (4.6) by performing the following manipulations. Take $t=\tilde{t},x=\tilde{x}$, $f_i=\tilde{f}_i$, replace w^i by $\tilde{t}+z_i$ and pass then from the functions to their respective Taylor-expansions at $z=0$.

We add two lemmas which later will be needed for technical purposes.

Lemma 8.2 If the n-th component f^n of f (cf. (7.1)) is constant, then the n-th component of $C(t,x,z_1,\dots,z_N;\ \mathcal{U}_o,\ \mathcal{U}_1,\dots,\ \mathcal{U}_N)$ is equal to x^n (= n-th component of x).

Proof. If N=1, then

$$(8.11)\quad C(t,x,z_1;\ \mathcal{U}_o,\ \mathcal{U}_1) = x + \sum_{\nu=1}^{\infty} \frac{1}{\nu!} h^{(\nu)}(t,x;\ \mathcal{U}_o,\ \mathcal{U}_1)z_1^{\nu}$$

(cf. Theorem 7.1). It can be seen immediately from (7.17) and (7.16) that - because of the hypothesis of the lemma - the n-th component of $h^{(\nu)}$ is zero, for $\nu = 1,2,\dots$. This proves the lemma in case N=1.

We proceed by induction with respect to N and use (8.8) for M = N-1. The expression on the right hand side of this formula is of the form (8.11), hence the n-th component is equal to the n-th component of x'. By hypothesis of induction however the letter turns out to be x^n.

The situation of Lemma 8.2 occurs if one considers along with the given system (7.1) an augmented system

$$(8.12)\quad \dot{x} = f(t,x;u),\quad \dot{x}^{n+1} = 1.$$

Here x^{n+1} is an additional state variabel. If we put $x^*=(x,x^{n+1})^T$, $f^*(t,x^*,u) = (f(t,x,u),0)^T$ we can associa~ with the system (8.12) the formal Taylor series

$$(8.13)\quad C^*(t,x^*,z_1,\dots,z_N;\ \mathcal{U}_o,\ \mathcal{U}_1,\dots,\ \mathcal{U}_N)$$

which is a (n+1)-dimensional vector. The last component of f^* is equal to 1,so the assertion of Lemma 8.2 holds for C^* (with x^{n+1} playing the role of x^n). But the special form (8.12) of the system in question gives rise to a stronger statement.

Lemma 8.3 We have

$$C^*(t,x^*,z_1,\dots,z_N;\ \mathcal{U}_o,\dots,\ \mathcal{U}_N) = \left(C(t,x,z_1,\dots,z_N;\mathcal{U}_o,\dots,\mathcal{U}_N),x^{n+1}\right)^T,$$ i.e. the first n components of C^* coincide with C, the last component is equal to x^{n+1}.

<u>Proof.</u> The case $N = 1$ is handled as in Lemma 8.2. In order to treat the general case we again use Theorem 8.1 with $M = N-1$. We apply the formula (8.8) to the function C^* and obtain

(8.14) $$C^*(t,x^*,z_1,\ldots,z_N;\ \mathcal{U}_o,\ \mathcal{U}_1,\ldots,\ \mathcal{U}_N) = \\ = C^*(t,\hat{x},z_N;\ \mathcal{U}_{N-1},\ \mathcal{U}_N)$$

where

$$\hat{x} = C^*(t,x^*,z_1,\ldots,z_{N-1};\ \mathcal{U}_o,\ldots,\ \mathcal{U}_{N-1}).$$

By hypothesis of induction the expression for $\hat{x}$ can be written in the form

$$\left(C(t,x,z_1,\ldots,z_{N-1};\ \mathcal{U}_o,\ldots,\ \mathcal{U}_{N-1}),\ x^{n+1}\right)^T .$$

Also, from what we already know in the case $N = 1$, the expression on the right hand side of (8.14) has the same $(n+1)$-th component as $\hat{x}$ and otherwise coincides with $C(t,x',z_N;\mathcal{U}_{N-1},\mathcal{U}_N)$, where x' consists of the first n components of $\hat{x}$. Using Theorem 8.1 once more we finally arrive at the desired result: The first n components of the vector on the right hand side of (8.14) can be identified with $C(t,x,z_1,\ldots,z_N;\ \mathcal{U}_o,\ldots,\ \mathcal{U}_N)$.

In the discussion of singular extremals (part II of this paper) certains sets of n-dimensional vectors play an important role. These sets will be denoted by the symbol $\mathcal{P}$. They can be derived by a formal procedure from the power series C which we just have introduced. So we conclude this section by giving the definition of these sets. This definition involves certain power series in a scalar variable λ which can be obtained from (8.2) by means of a substitution of the form

(8.15) $$z \to z(\lambda)\ ,\ \mathcal{U}_o \to \mathcal{U},\ \mathcal{U}_i \to \mathcal{U}_i(\lambda), i=1,\ldots,N .$$

Here $z(\lambda) = (z_1(\lambda),\ldots,z_N(\lambda))^T$ is supposed to be a N-tupel of scalar functions of λ which are of class C^∞ on some interval $[0,\lambda_o]$, $\lambda_o > 0$, and which satisfy the following conditions (for $\lambda \in [0,\lambda_o]$)

(8.16) $$z(0) = 0,\quad z_1(\lambda) \le z_2(\lambda) \le \ldots \le z_N(\lambda) \le 0.$$

$\mathbf{U}_i(\lambda)$ is a sequence of the form (7.8) which satisfies this condition (for $i=1,\dots,N$).

(8.17) $u_{i,\rho} = u_{i,\rho}(\lambda)$ is a polynomial in λ. It vanishes identically for almost all ρ.

Because of the first of the relations (8.16) the Taylor-expansion of $\frac{1}{\underline{\nu}!}K_{\underline{\nu}}(t,x;\ \mathbf{U},\ \mathbf{U}_1(\lambda),\dots,\ \mathbf{U}_N(\lambda))z(\lambda)^{\underline{\nu}}$ at $\lambda = 0$ does not contain terms λ^{σ} for $\sigma < |\underline{\nu}|$. Hence the formal sum of all these Taylor-expansions makes sense and leads to a formal power series in λ (which can be obtained in the usual way by collecting equal powers of λ). This power series will then be denoted by

(8.18) $C(t,x,z_1(\lambda),\dots,z_N(\lambda);\ \mathbf{U},\ \mathbf{U}_1(\lambda),\dots,\ \mathbf{U}_N(\lambda))$.

Definition 8.1. Let U be a subset of the u-space, and let there be given a scalar t, a n-dimensional vector x and a sequence $\mathbf{U} = \{u_o,u_1,\dots\ \dots\}$ which satisfy these conditions

(8.19) $(t,x,u_o) \in Y, \quad u_o \in U$

We then denote by $\mathbf{P} = \mathbf{P}_U(t,x,\mathbf{U})$ the collection of all n-dimensional vectors p which have the following property. One can find two positive integers N and s, a N-dimensional vector $z(\lambda)$ and sequences $\mathbf{U}_i(\lambda)$ such that the following conditions are satisfied

(i) $z(\lambda)$ is of class C^{∞} on some interval $[0,\lambda_o]$, (8.16) holds,

(ii) $(t,x,u_{i,o}(0))\in Y, \quad u_{i,o}(0) \in \operatorname{int}U$, (8.17) holds, $i=1,\dots,N$.

(iii) the formal power series of the form (8.18) which is associated with $z(\lambda),\ \mathbf{U},\ \mathbf{U}_1(\lambda),\dots,\ \mathbf{U}_N(\lambda)$ **is of the form**

$$x + \lambda^s p + \mathcal{O}(\lambda^{s+1}).$$

We wish to point out, that N,s and of course $z(\lambda)$, $\mathbf{U}_i(\lambda)$ may **vary with** $p \in \mathbf{P}$.

We add three remarks, to which we have to refer at later occasions.

Remark 1. As one can see from our previous considerations the definition of the power series (8.18) makes sense also in case the $u_{i,\rho}$ are arbitrary formal power series in λ. However admitting these choices of $\mathcal{U}_i(\lambda)$ in the above definition ($u_{i,o}(0)$ of course has then to be understood as the λ-free term in $u_{i,o}(\lambda)$) will not yield a larger set $\mathcal{P}$. This follows simply from the fact that contributions to the coefficient of some power λ^{μ} in (8.18) can arise only from $K_{\underline{\nu}}$ with $|\underline{\nu}| \leq \mu$. The latter however depend upon $u_{i,j}$ with $j < \mu$ only (Lemma 7.2). In addition it is clear that there will be also no contribution from those terms in the series representation of $u_{i,j}$ which contain powers λ^{σ} for $\sigma > \mu$. Hence the coefficients of λ^{τ} for $\tau \leq s$ in the formal series (8.18) remain unchanged if all $u_{i,\rho}$ with $\rho > \mu$ are replaced by 0 and the remaining power series $u_{i,j}$ are replaced by suitable partial sums.

Remark 2. $p \in \mathcal{P}$ implies $\sigma p \in \mathcal{P}$ for any real number $\sigma \geq 0$. This can simply be put into evidence by means of the substitution $z \to \sigma^{1/s} z$.

Remark 3. Consider the augmented system (8.12) and let $\mathcal{P}^* = \mathcal{P}_U(t,x^*,u)$, $x^* = (x,x^{n+1})^T$ be the set (in R^{n+1}) associated with it according to Definition 8.1. Then any $p^* \in \mathcal{P}^*$ can be written in the form $(p,0)^T$ with $p \in \mathcal{P}$. This follows directly from Lemma 8.3.

Remark 4 if $u(\cdot)$ is an admissible control function and $x(\cdot)$ the corresponding trajectory then the set $\mathcal{P}_U(t,x(t),\mathcal{U}(t))$, ($\mathcal{U}(t)$ being the sequence (7.5)) is well defined for every t which is such that $u(t) \in \text{int } U$. It is this set which actually will be studied in part II of this work. It does however not provide a suitable setting for the multiplier rule to be discussed in the next section, since this rule, if it is confined to the sets $\mathcal{P}_U$ only, does not comprise the maximum principle. For this reason a somehow more general set is associated with the points of the reference solution.

9. Control Constraints. Higher Order Necessary Conditions.

We now include control constraints into our considerations. So we assume that there is given, in addition to the data described in the beginning of Sec. 7, a non-empty subset U of the u-space. We call U the control set. By an admissible control function $u(\cdot)$ we mean a function which has the following two properties.

(9.1) (i) $u(t) = u(t-0) \in U$ for every t. (ii) To each $\tilde{t}$ there exists $\varepsilon > 0$ and a function $\tilde{u}(t)$, which is of class C^{∞} on $(\tilde{t}-\varepsilon, \tilde{t}+\varepsilon)$ and satisfies $u(t) = \tilde{u}(t)$ for $\tilde{t}-\varepsilon \le t < \tilde{t}$.

It follows from this definition that $u(\cdot)$ and all its derivatives are piecewise continuous. For notational convenience we agree to identify the value of a function and its derivatives with the respective left hand limits. The sequence (7.5) is therefore well defined for every t whenever $u(\cdot)$ is an admissible control function.
Let there be given an admissible control function $u(\cdot)$ and a solution $x(\cdot)$ of the diff.eq.

$$\dot{x} = f(t,x,u(t))$$

which satisfies the condition

$$(t,x(t),\ u(t\pm 0)) \in Y$$

for all t in some interval $I = [t_o, t_e]$. $u(\cdot)$ and $x(\cdot)$ will be called reference control and reference trajectory respectively, the pair $(u(\cdot),x(\cdot))$ will be called reference pair. All these data have to be regarded as fixed throughout this section. By $\mathbf{U}(t)$ we will denote the sequence of the form (7.5) which is associated with the ref. control for a given t.
The main objective of this section is the presentation of higher order necessary conditions for optimal solutions. The statement of these conditions will involve certain subsets of $\mathbb{R}^n$ which we associate to the given reference pair and to every real number $t \in [t_o, t_e]$. These sets will be denoted by Π_t. The precise definition will be given in two steps. We first introduce the notion of a control variation.

<u>Definition 9.1</u> A family of bounded admissible control functions $u(\cdot,\lambda)$ depending upon a scalar parameter λ, $0<\lambda\le\lambda_o$, is called a "control variation concentrated at $\tilde{t}$" if it satisfies the following two conditions.

(9.2) (i) $(t,x(t),u(t\pm 0,\lambda))\in Y,\ t_o\le t\le t_e,\ 0<\lambda\le\lambda_o,$

(ii) $u(t,\lambda) = u(t)$ (=ref. control) for $t\notin[\omega(\lambda),\tilde{t}]$ where $\omega(\lambda)\le\tilde{t}$ and $\lim_{\lambda\to 0}\omega(\lambda) = \tilde{t}$.

Before we proceed to the next definition we wish to point out that the initial value problem

(9.3) $\dot{x} = f(t,x;u(t,\lambda)),\quad x(t_o) = a$

admits a solution $x(t;\lambda,a)$ which exists on $[t_o,t_e]$ and satisfies the inequality

(9.4) $(t,x(t;\lambda,a),u(t\pm 0,\lambda))\in Y,\ t_o\le t\le t_e$

provided λ is sufficiently small and a sufficiently close to $x(t_o)$ (= initial value of the reference trajectory). This follows by standard arguments from the properties (9.2) of a control variation.

<u>Definition 9.2</u> Given a reference pair $u(\cdot)$, $x(\cdot)$ on the interval $[t_o,t_e]$ and let $x_o = x(t_o)$ be the initial value of the trajectory. For any $\tilde{t}\in(t_o,t_e)$ we denote by $\Pi^*_{\tilde{t}}$ the collection of all n-dimensional vectors p having the following property. There exists a control variation concentrated at $\tilde{t}$ such that the solution $x(\cdot;\lambda,a)$ of the initial value problem (9.3) is - for the given fixed $\tilde{t}$ - a C^∞-function of λ,a on a set of the form

(9.5) $\{\lambda,a : 0\le\lambda\le\lambda_o,\ \|a-x_o\|\le\varepsilon\}$

and admits the asymptotic expansion

(9.6) $x(\tilde{t};\lambda,x_o) = x(\tilde{t}) + \lambda^s p + \mathcal{O}(\lambda^{s+1})$,

s being some positive integer (which may depend upon p).

Furthermore, for any $\tilde{t}\in[t_o,t_e]$, we denote by $\Pi_{\tilde{t}}$ the collection of all $\tilde{p}$ having the following property. There exists a subintervall $\mathfrak{J}$ of $[t_o,t_e]$ - which may depend upon $\tilde{t}$ - and a continuous function $p(\cdot)$ on $\bar{\mathfrak{J}}$ such that $\tilde{t}\in\bar{\mathfrak{J}}$, $p(\tilde{t}) = \tilde{p}$ and $p(t)\in\Pi^*_t$ for all $t\in\mathfrak{J}$.

<u>Remarks.</u> (i) The definition of Π^*_t remains unchanged if the initial time t_o is replaced by any t_o' with $t_o\le t_o'<\tilde{t}$ and x_o is replaced by the value $x(t_o')$ of the reference trajectory at $t=t_o'$. Also the exponent s appearing in (9.6) may be replaced by any multiple rs, $r\ge 1$ and integer, without changing p. This is clear since the substitution $\lambda\to\lambda^r$ in the control variation $u(\cdot,\lambda)$ will change the corresponding trajectory to $x(\cdot;\lambda^r,a)$.

(ii) If $p\in \Pi^*_t$ $(p\in \Pi_t)$ then also $\alpha p\in \Pi^*_t$ $(\alpha p\in \Pi_t)$ for any positive scalar α. This is clear, since the transformation $\lambda \to \alpha^{1/s}\lambda$ will make αp the leading coefficient in the asymptotic expansion (9.6).

(iii) We have $\mathbf{P}_U(\tilde{t},x(\tilde{t}),\mathcal{U}(\tilde{t}))\subseteq \Pi^*_{\tilde{t}}$ for every $\tilde{t}\in(t_o,t_e)$. In other words:

$$p\in \mathbf{P}_U(\tilde{t},x(\tilde{t}),\mathcal{U}(\tilde{t})) \quad \text{implies} \quad p\in \Pi^*_{\tilde{t}}\,.$$

This can be verified by a careful comparison of the definitions given in this section and Definition 8.1. To begin with we remark that the sequences $\mathcal{U}_i(\lambda)$ which appear in the latter one can be represented in the form

$$\{u_i(\tilde{t},\lambda),\ (\partial u_i/\partial t)(\tilde{t},\lambda),\qquad (\partial^2 u_i/\partial t^2)(\tilde{t},\lambda),\ldots\}\ ,$$

where $u_i(t,\lambda) := u(t,\tilde{t}\ \mathcal{U}_i(\lambda))$ and the control functions appearing on the right hand side are defined according to (7.6) (with $\mathcal{U}_i(\lambda)$ instead of $\mathcal{U}$ and $\tilde{t}$ instead of τ). It follows from this formula and from condition (ii) of Definition 8.1 that

$$u_i(t,\lambda)\in U,\quad (t,x(t),u_i(t,\lambda))\in Y$$

provided λ and $|t-\tilde{t}|$ are sufficiently small.

Given now sequences $\mathcal{U}_i(\lambda)$ and a N-dimensional vector $z(\lambda)$ as specified in Definition 8.1. A control variation concentrated at $\tilde{t}$ is then constructed by means of the following procedure:

Replace $u_o(\cdot)$ by $u(\cdot)$, $u_i(\cdot)$ by $u_i(\cdot,\lambda)$ on the right hand side and z by $z(\lambda)$ on the left hand side of the definition (8.6). Call the resulting function $u_{z(\lambda)}(t,\lambda)$ and consider the initial value problem

$$\dot{x} = f(t,x;u_{z(\lambda)}(t,\lambda)),\quad x(t_o) = a$$

and denote its solution by $x(t;\lambda,a)$. It is easy to see that the above differential equation can also be written in the form (2.13) with $p=\lambda$, $w=w(\tilde{t})+z(\lambda) =: w(\lambda)$. The function $f(t,x,\lambda|w(\lambda))$ appearing on the right hand side admits a representation of the form (2.3) with

(9.7) $\quad f_o(t,x,\lambda) = f(t,x;u(t)),f_i(t,x,\lambda) = f(t,x;u_i(t,\lambda))\quad$ for $\ i>0$.

It follows therefore from Lemma 2.1 - or rather from its extension to the case of parameter - dependent equations which was discussed in connection with (2.13) - that $x(\tilde{t};\lambda,a)$ is a C^∞function of λ,a on a set of the form (9.5). Furthermore the identity

$$x(\tilde{t};\lambda,x_o) = c(\tilde{t},x;w(\tilde{t})+z(\lambda)|\quad f_o(\cdot,\lambda),\ldots,f_N(\cdot,\lambda)),x=x(\tilde{t}),$$

holds true, in view of our previous remarks concerning the solution of the initial value problem (8.7). We can now apply Lemma 8.1 and arrive then at this result: The Taylor-expansion (at $\lambda = 0$) of $x(\tilde{t};\lambda,x_o)$ is given by the formal power series

$$C(\tilde{t},x,z_1(\lambda),\ldots,z_N(\lambda),\ \mathcal{U}_o(\tilde{t},\lambda),\ldots,\mathcal{U}_N(\tilde{t},\lambda)),$$

where the sequences $\mathcal{U}_i(\tilde{t},\lambda)$ consist just of the time-derivatives (at $t=\tilde{t}$) of the control functions $u_o(t) = u(t)$ and $u_i(t,\lambda) = u(t,\tilde{t};\ \mathcal{U}_i(\lambda))$ (cf. (7.7)) respectively. It follows however from Lemma 7.1 that these derivatives coincide with $\mathcal{U}(t)$ and $\mathcal{U}_i(\lambda)$ respectively and the Taylor-expansion of $x(\tilde{t};\lambda,x_o)$ finally turns out to be nothing else than the formal power series (8.18) for $\mathcal{U} = \mathcal{U}(\tilde{t})$. Thereby it has been shown that any $p \in P_U(\tilde{t},x(\tilde{t}),\ \mathcal{U}(\tilde{t}))$ appears as leading coefficient in the Taylor-expansion of some $x(\tilde{t};\lambda,x_o)$ which belongs to a suitable control variation concentrated at $\tilde{t}$. Hence we have indeed $p \in \Pi^*_{\tilde{t}}$.

(iv) The relation

$$(9.8)\qquad f(t,x(t);u) - f(t,x(t);u(t)) \in \Pi_t$$

holds whenever $t \in (t_o,t_e], u \in U,\ (t,x(t),u) \in Y$. This statement follows by a straightforward continuity argument once we know that

$$(9.8')\qquad f(\tilde{t},x(\tilde{t});u) - f(\tilde{t},x(\tilde{t});u(\tilde{t})) \in \Pi^*_{\tilde{t}}$$

for every pair $(\tilde{t},u)$ which satisfies the condition

$$\tilde{t} \in (t_o,t_e),\ u \in U,\ (\tilde{t},x(\tilde{t}),u) \in Y.$$

Regard $\tilde{t},u$ as fixed and consider the control variation which is defined as

$$u(t,\lambda) = \begin{cases} u(t) & \text{for } t \le \tilde{t}-\lambda \text{ and } t > \tilde{t}, \\ u & \text{für } \tilde{t}-\lambda < t \le \tilde{t}. \end{cases}$$

It is immediately verified that for this choice of $u(t,\lambda)$ the solution $x(t;\lambda,a)$ of the initial value problem (9.3) is a C^∞-function of λ,a (cf. e.g. Lemma 2.1). That (9.8') holds true is then a consequence of Theorem 3.1. Indeed the vector appearing in (9.8') is nothing else than the derivative $(\partial x/\partial\lambda)(\tilde{t};\lambda,a)$ at $\lambda=0$, $a = x_o$.

We now turn to the consideration of optimal solutions and assume that the reference trajectory satisfies boundary conditions of the form

$$(9.9)\qquad x(t_o) \in M_o,\quad x(t_e) \in M_e,$$

where M_o,M_e are given subsets of the state space and are defined in terms of equations and inequalities. Furthermore we assume that the reference pair minimizes a certain integral in comparison with all admissible trajectories starting at M_o at $t=t_o$ and reaching M_e at $t=t_e$. As usual we think of the value of the integral as being identical with the terminal component of the (augmented) state variable. Furthermore we assume that the definitions of M_o, M_e are extended to the (augmented) state space in such a way that both the constraints (9.9) and the optimality of the reference pair can be accounted for by the transversality conditions. Here and in the sequel we will use the notion "transversality condition" and "regular point" in the usual sense (see e.g. [1], Ch.VI, Sec.7 and Ch.VII, Sec. 3).

Theorem 9.1. (Higher Order Necessary Conditions= HNC).
Let the reference pair be optimal and let $x(t_o), x(t_e)$ respectively be regular points of M_o, M_e respectively. Then there exists a non-trivial solution $y(\cdot)$ of the adjoint variational equation (taken along the reference pair) which satisfies the transversality conditions at $t=t_o, t_e$ and the inequalities $y(t)^T p \leq 0$ for all $p \in \Pi_t$ and all $t \in [t_o, t_e]$.

The proof of the theorem will be given in Sec. 11.

We conclude this section with two remarks concerning special cases of the HNC.

As an immediate consequence of the last remark following Definition 9.2 we have the inequality

$$y(t)^T f(t,x(t);u) \leq y(t)^T f(t,x(t);u(t))$$

whenever $u \in U$ and $(t,x(t),u) \in Y$. This means of course that the Maximum Principle is included in the HNC.

Also the "high order maximal principle" of Krener [3] is a consequence of Theorem 9.1. In order to put this into evidence let us consider a "control variation $\alpha(\lambda)x$ to $u^o(t)$" as defined in [3], Sec. 2. We adopt for the moment the notation used in [3] except that we keep our present custom to index variations by λ instead of s. Note also that in our statement of higher oder conditions the terminal time is regarded to be fixed. Let $\alpha(\lambda)$ now be decomposed into variations $\gamma^i(p_i(\lambda))$ and $\gamma^i(q_j(\lambda))$ according to formula (2.4) in [3]. Since the terminal time is fixed the polynomials p_i, q_j have to satisfy

$$(9.10) \qquad q_1(\lambda) + q_2(\lambda) + \sum_{i=1}^{k} p_i(\lambda) = 0$$

for all λ. Using the definition of $\alpha(\lambda)$ and $\gamma^i(\ldots)$ as given in

[3] it is not difficult to confirm the following statements. For every fixed $\tilde{t}$ and $\lambda>0$ one can identify the vector

$$(9.11) \qquad \gamma^{o}(-q_2(\lambda))\alpha(\lambda)x(\tilde{t}+q_2(\lambda))$$

with the value at $t=\tilde{t}$ of a solution of equation (9.3) The corresponding admissible control $u(\cdot,\lambda)$ coincides with the reference control for $t\leq\tilde{t}+q_1(\lambda)+q_2(\lambda)$. In other words, $u(\cdot,\lambda)$ is a control variation concentrated at $t=\tilde{t}$. Furthermore the vector (9.11), regarded as a function of λ, is of class C^{∞} in a full neighborhood of $\lambda=0$. The first non-vanishing coefficient in the Taylor-expansion of (9.11) at $\lambda=0$ therefore belongs to the set Π^{*}_{t}.

Now $\alpha(\lambda)x$ is said to be of order h at $x=x(\tilde{t})$ if the $h-1$ first derivatives with respect to λ of $\alpha(\lambda)x(t)$ vanish identically in t, for all t in a neighborhood of $\tilde{t}$. This means that the asymptotic relation

$$(9.12) \qquad \alpha(\lambda)x(t) = x(t) + \frac{1}{h!}\lambda^{h}p(t) + \mathcal{O}(\lambda^{h+1})$$

holds uniformly in t where

$$(9.13) \qquad p(t) = \frac{d^{h}}{d\lambda^{h}}\alpha(0)x(t).$$

We wish to demonstrate that (9.12) implies

$$(9.14) \qquad p(\tilde{t})\in\Pi_{\tilde{t}} .$$

Once this has been done it is clear - in view of (9.13) - that the high order maximal principle is indeed included in Theorem 9.1.

Using the assumption $q_i(0) = p_i(0) = 0$ one derives from (9.12) a further asymptotic relation, namely

$$(9.15) \qquad \alpha(\lambda)x(\tilde{t}+q_2(\lambda)) = x(\tilde{t}+q_2(\lambda)) + \frac{1}{h!}\lambda^{h}p(\tilde{t}) + \mathcal{O}(\lambda^{h+1}) .$$

In order to obtain the Taylor-expansion of (9.11) one has to apply the transformation $\gamma^{o}(-q_2(\lambda))$ to the right hand side of (9.15). The result can be easily evaluated in view of the formula

$$(9.16) \qquad \gamma^{o}(\tau)(x(t)+z) = x(t+\tau)+z+\mathcal{O}(\|z\tau\|) + \sigma(\|z\|).$$

The validity of (9.16) can be inferred by standard arguments if one assumes (as is done in [3]) that the underlying system does not depend explicitly upon t. Combining (9.15) with (9.16) one finally sees that the Taylor-expansion of (9.11) starts with the term

$$x(\tilde{t}) + (h!)^{-1} p(\tilde{t})\lambda^h .$$

Thereby we have proved that

(9.14') $\quad p(\tilde{t}) \in \prod_{\tilde{t}}^{*}.$

The same result holds if $\tilde{t}$ is replaced by any t sufficiently close to $\tilde{t}$. Since $p(\cdot)$ is a piecewise continuous function of t it is clear that one can pass from (9.14') to the desired result (9.14).

10. Auxiliary results.

Lemma 10.1. Let $\varphi = \varphi(\xi_1,\ldots,\xi_s)$ be a scalar function of the scalar variables $\xi_1,\ldots,\xi_s$ and let there be given s scalar functions $\xi_i(\eta) = \xi_i(\eta_1,\ldots,\eta_r)$ of the scalar variables $\eta_1,\ldots,\eta_r$. Assume that all functions are of class C^∞ (though differentiability of sufficient high order would suffice). Let $\chi(\cdot)$ be the composite function which is given as

$$\chi(\eta) := \varphi(\xi_1(\eta),\ldots,\xi_s(\eta)) .$$

Claim: Every partial derivative

$$\partial^\sigma \chi/(\partial\eta_1)^{\mu_1}\ldots\partial(\eta_r)^{\mu_r} \quad , \ \sigma = \mu_1+\mu_2+\ldots+\mu_r \geq 1, \quad \text{can be written as a}$$

linear combination of power-products

$$\zeta_1^{i_1}\ldots\zeta_k^{i_k} \quad , \ (i_1,\ldots,i_k) \neq (0,\ldots,0),$$

where the ζ_τ are the -somehow ordered- partial derivatives of the form

$$\partial^{\sigma'}\xi_i/(\partial\eta_1)^{\nu_1}\ldots.(\partial\eta_r)^{\nu_r}, \nu_j \leq \mu_j, j=1,\ldots,r, i=1,\ldots,s$$

and the coefficients are derivatives of φ.

Proof. We have, by the chain rule,

$$\partial\chi/\partial\eta_j = \sum_{i=1}^{s} (\partial\varphi/\partial\xi_i)(\partial\xi_i/\partial\eta_j),$$

and this proves the statement in case $\sigma=1$. The general case follows by induction and repeated application of the chain rule. Thereby the lemma is proved.

As in the previous section we assume that a fixed reference pair $u(\cdot), \tilde{x}(\cdot)$*) is given on the interval $[t_o, t_e]$. We introduce the transition matrix $\phi(t,\tau)$ of the variational equation along the ref. pair. ϕ is uniquely defined by these relations

(10.1) $\partial\phi/\partial t = f_x(t, \tilde{x}(t); u(t))\phi, \quad \phi(\tau,\tau) = I$ (= id.matrix).

The proof of the separation properties of the cone of attainability to be defined in the next section is based on the existence of certain smooth families of admissible pairs.
The next lemma and its corollary provides the essential tool for the construction of such families.

Lemma 10.2 Let there be given finitely many, say k, real numbers t_j and for each $j=1,\dots,k$ a n-dimensional vector p_j such that

(10.2) $t_o < t_1 < t_2 < \dots < t_k < t_e, \quad p_j \in \Pi^*_{t_j}$.

Let $\hat{\tau} = (\tau_1, \dots, \tau_k)^T$ denote a k-dimensional parameter restricted to the set

(10.3.) $\{\hat{\tau} = (\tau_1, \dots, \tau_k)^T : 0 \le \tau_j \le \varepsilon, \ j=1,\dots,k\}$.

Claim: One can find $\varepsilon > 0$, a positive integer s and a m-dimensional vector $u(t,\hat{\tau})$ which is defined for all t and all $\hat{\tau}$ satisfying (10.3) and which has the following properties

(i) $u(\cdot,\hat{\tau})$ is an admissible control function for each $\hat{\tau}$.

(ii) The solution $x(t;\hat{\tau},a)$ of the initial value problem

(10.4) $\dot{x} = f(t,x;u(t,\hat{\tau})), \quad x(t_o;\hat{\tau},a) = a = (a_1,\dots,a_n)^T$

exists on $[t_o,t_e]$ whenever $(\hat{\tau} = (\tau_1,\dots,\tau_k)^T, a)$ belongs to the set

(10.5) $\mathcal{T} = \{\hat{\tau}, a : 0 \le \tau_j \le \varepsilon, j=1,\dots,k, \ \|a - \tilde{x}_o\| \le \varepsilon\}$

($\tilde{x}_o$ = initial value of the ref. trajectory).

(iii) $x(t;0,\tilde{x}_o) = \tilde{x}(t)$ (=ref. trajectory).

(iv) $x(t_e;\hat{\tau},a) =: x_e(\hat{\tau},a)$ is a C^∞-function of $\hat{\tau}, a$ on $\mathcal{T}$. At $(\hat{\tau},a) = (0,\tilde{x}_o)$ the Jacobian matrix $\partial x_e/\partial a$ coincides with $\phi(t_e,t_o)$.

---

*) In order to avoid notational difficulties we denote in this section the reference trajectory by $\tilde{x}(\cdot)$ (and not by $x(\cdot)$) .

(v) At $(\hat{\tau},a) = (0,\tilde{x}_o)$ the derivatives with respect to the components of $\hat{\tau}$ satisfy these conditions

(10.6)
$$\frac{\partial^{\sigma_1+\dots+\sigma_k} x_e}{\partial\tau_1^{\sigma_1}\dots\partial\tau_k^{\sigma_k}} = 0 \quad \text{if} \quad \sum_j \sigma_j \le s \quad \text{and} \quad \max_j \sigma_j < s$$

$$\frac{\partial^s x_e}{\partial\tau_i^s} = s!\ \phi(t_e,t_i)p_i,\ i=1,\dots,k\ .$$

Proof. For every $j=1,\dots,k$ there exists a control variation $u_j(\cdot,\lambda)$, $0\le\lambda\le\lambda_o$, and a corresponding family of trajectories $x_j(\cdot;\lambda,a)$ such that

(10.7)
$$x_j(t_{j-1};\lambda,a) = a,$$
$$x_j(t_j;\lambda,\tilde{x}(t_{j-1})) = \tilde{x}(t_j) + \lambda^s p_j + \mathcal{O}(\lambda^{s+1}).$$

If λ_o is sufficiently small then $u_j(t,\lambda) = u(t)$ for $0\le\lambda\le\lambda_o$ and $t\notin[t_{j-1},t_j]$. Furthermore $x_j(t;0,a)$ is solution of the initial value problem

(10.8) $\dot{x} = f(t,x;u(t))$, $x(t_{j-1}) = a$

on $[t_{j-1},t_j]$ and therefore we have in particular

$$x_j(t;0,\tilde{x}(t_{j-1})) = \tilde{x}(t) \text{ if } t_{j-1}\le t\le t_j$$

$x_j(t_j;\lambda,a)$ finally is a C^∞-function of λ,a provided λ and $||a-\tilde{x}_o||$ are sufficiently small. All this follows from Definition 9.2 and the subsequent remarks. Note in particular that one can assume without loss of generality that the exponent s in the asymptotic relation (10.7) is independent from j. This need not be true from the beginning, however one can always replace the individual exponents by their least common multiple. Since the solution of the initial value problem (10.8) assumes the value $x_j(t_j;0,a)$ at $t=t_j$ it is clear that

(10.9) $\phi(t_j,t_{j-1}) = \partial x_j(t_j;0,a)/\partial a$ for $a=\tilde{x}(t_{j-1})$.

We put $t_{k+1} = t_e$ and introduce $x_{k+1}(t;a)$ as the solution of the initial value problem (10.8) for $j=k+1$. Then we define $x(t;\hat{\tau},a) = x(t;\tau_1,\dots,\tau_k,a)$ recursively as follows

(10.10)
$$x(t_o;\hat{\tau},a) = a$$
$$x(t;\hat{\tau},a) = x_j(t;\tau_j,x(t_{j-1};\hat{\tau},a)) \text{ for } t_{j-1}<t\le t_j \text{ and } j=1,\dots,k+1\ .$$

This definition makes sense if $t \in [t_o, t_e]$ and $(\hat{\tau}, a) \in \mathcal{T}$, provided the bound ε in (10.5) is sufficiently small. Proceeding by induction with respect to j and using (10.7), (10.9) one easily verifies the following statements.

(i) $x(t;\hat{\tau},a)$ is for fixed $\hat{\tau}, a$ an admissible trajectory on the interval $[t_o, t_j]$ and we have $x(t;0,\tilde{x}_o) = \tilde{x}(t)$ for all $t \in [t_o, t_j]$.

(ii) $x(t_j;\hat{\tau},a)$ is of class C^∞ on the set $\mathcal{T}$ (cf. (10.5)) and depends upon the components $\tau_1, \ldots, \tau_j$ of $\hat{\tau}$ only.

(iii) The partial derivatives of $x_j = x(t_j;\hat{\tau},a)$ at $\hat{\tau} = 0$, $a = \tilde{x}_o$ satisfy the conditions

$$\partial x_j / \partial a = \phi(t_j, t_o),$$

$$\frac{\partial^{\sigma_1 + \ldots + \sigma_k} x_j}{\partial \tau_1^{\sigma_1} \ldots \partial \tau_k^{\sigma_k}} = 0 \quad \text{if} \quad \sum_i \sigma_i \leq s \text{ and } \max_i \sigma_i < s \quad \text{(cf. Lemma 10.1)},$$

$$\frac{\partial^s x_j}{\partial \tau_i^s} = s!\,\phi(t_j, t_i) p_i \quad \text{if} \quad i \leq j.$$

Thereby the lemma is proved since the last statement for $j=k+1$ gives the desired result.

<u>Corollary.</u> Let the hypothesis of the lemma be satisfied. Then there exists a function $x(\tau_1, \ldots, \tau_k, a) = x(\hat{\tau}, a)$ which is defined and continous on the set $\mathcal{T}$ for ε sufficiently small (cf. (10.5)) and which has the following properties.

(i) $x(\hat{\tau}, a)$ is terminal point for an admissible trajectory starting at $x=a$ for $t=t_o$.

(ii) We have the asymptotic expansion

$$x(\hat{\tau}, a) = \tilde{x}(t_e) + \phi(t_e, t_o)(a - \tilde{x}_o) + \sum_{j=1}^{k} \tau_j \phi(t_e, t_j) p_j + r(\hat{\tau}, a),$$

where

$$\lim_{(\hat{\tau}, a) \to (0, \tilde{x}_o)} \frac{r(\hat{\tau}, a)}{\|\hat{\tau}\| + \|a - \tilde{x}_o\|} = 0.$$

Proof. A function having the properties (i) and (ii) can be obtained from $x_e(\hat{\tau},a)$ (cf. the statement of the lemma) by means of the substitution $\tau_j \to (\tau_j)^{1/s}$, $j=1,\dots,k$.

The statement of the corollary can be rephrased using the notion of a differential cone (see [10], Ch.4, Sec.2). This version will actually be applied later.

11. Proof of Theorem 9.1.

Lemma 11.1. Let $u(\cdot)$, $\tilde{x}(\cdot)$ be a reference pair (not neccessarily optimal) and let Π_t, $t\in[t_o,t_e]$ be the sets associated with it according to Definition 9.2. Then one can find a countable sequence of pairs (t_ν,p_ν), $\nu=1,2,\dots,$ with the following properties.

(i) $t_\nu \in (t_o,t_e)$, $t_\nu \neq t_\mu$ if $\nu\neq\mu$, $p_\nu \in \Pi_{t_\nu}$,

(ii) the vectors $\phi(t_e,t_\nu)p_\nu$, $\nu=1,2,\dots$ form a dense subset of the collection of all vectors of the form

(11.1) $\phi(t_e,t)p$, $p\in\Pi_t$, $t\in[t_o,t_e]$.

Proof. Let $\phi(t_e,\tilde{t}_\nu)\tilde{p}_\nu$, $\tilde{p}_\nu\in\Pi_{\tilde{t}_\nu}$, $\tilde{t}_\nu\in[t_o,t_e]$,

be a sequence of vectors which is dense in the collection of all vectors of the form (11.1). For every $\nu= 1,2,\dots$ we choose an open non-empty subinterval $\mathfrak{J}_\nu$ of (t_o,t_e) and a function $p_\nu(t)$ which is continuous on $\bar{\mathfrak{J}}_\nu$ such that

$$\tilde{t}_\nu\in\bar{\mathfrak{J}}_\nu ,\quad p_\nu(t)\in\Pi_t \text{ for all } t\in\mathfrak{J}_\nu,\ \tilde{p}_\nu = p_\nu(\tilde{t}_\nu).$$

This is always possible according to Definition 9.2. It is now easy to see that one can select a sequence $t_{\nu,\mu}$ of real numbers in the open interval (t_o,t_e) such that these conditions are satisfied

$$t_{\nu,\mu} \neq t_{\nu',\mu'} \quad \text{if} \quad (\nu,\mu) \neq (\nu',\mu') ,$$

$$t_{\nu,\mu} \in\mathfrak{J}_\nu ,\quad |t_{\nu,\mu} - \tilde{t}_\nu|\le 1/\mu .$$

Since $\tilde{p}_\nu = \lim_{\mu\to\infty} p_\nu(t_{\nu,\mu})$ it is also clear that the pairs $(t_{\nu,\mu}, p_\nu(t_{\nu,\mu}))$ form a sequence which has the desired properties (i) and (ii).

For the remaining portion of this section we restrict initial and terminal values of the admissible trajectories to sets M_o, M_e which are given in terms of equations and inequalities. Hence M_o is a set of the form

(11.2) $M_o = \{x \in \mathbb{R}^n : \beta_i(x) = 0 \text{ for } i=1,\dots,r, \beta_i(x) \le 0 \text{ for } i=r+1,\dots,s\}$

We assume that the β_i are continuously differentiable functions of x in some neighborhood of $\tilde{x}(t_o)$ (= initial value of the ref. trajectory). The regularity requirement which is part of the condition (9.4) has the following precise meaning :

If the numeration of the β_i is rearranged in such a way that we have

(11.3)
$\beta_i(\tilde{x}(t_o)) = 0$ for $i=1,\dots,r'$, $\beta_i(\tilde{x}(t_o)) < 0$ for $i > r'$,

then the vectors $b_i = \operatorname{grad}\beta_i$ are linearly independent at $x = \tilde{x}(t_o)$, for $i=1,\dots,r$.

It follows from this condition that one can define in a natural way a tangent cone T_o to M_o at $\tilde{x}(t_o)$ (see e.g. [1], Ch. VII, Sec. 3).

The hypotheses concerning the terminal manifold M_e are analogous to (11.2) and (11.3), here however we have to include an inequality which accounts for the optimality of the ref. pair. If, e.g., the quantity to be minimized is the terminal value $x^1(t_e)$ of the first component of x then one has to assume that the inequality $x^1 - x^1(t_e) \le 0$ appears among the relations defining M_e. The tangent cone to M_e at $\tilde{x}(t_e)$ will be denoted by T_e. We now claim: If the ref. pair $u(\cdot), \tilde{x}(\cdot)$ is optimal (in the sense as explained in Theorem 9.1) and if (t_ν, p_ν) is a sequence of pairs as described in Lemma 11.1 then one can find a vetor $y_e \neq 0$ such that the following relations hold true

(11.4)
$$y_e^T k \ge 0 \text{ for } k \in T_e,\quad y_e^T \phi(t_e,t_o)c \le 0 \text{ for } c \in T_o,$$
$$y_e^T \phi(t_e,t_\nu)p_\nu \le 0 \quad \text{for } \nu=1,2,\dots .$$

Once the existence of such an y_e has been established the proof of Theorem 9.1 is straightforward. Indeed we have then, for reasons of continuity, $y_e^T \phi(t_e,t)p \le 0$, for all $p \in \Pi_t$ and all $t \in [t_o,t_e]$, and this of course is nothing else than the statement of the theorem, $y(t)$ being the solution of the adjoint variational equation with the terminal value $y(t_e) = y_e$.

We choose a function $g(x)$ which vanishes together with all partial derivatives at $x=0$ and is such that $\tilde{x}(t_o)+c+g(c)\in M_o$ whenever $c\in T_o$ and $\|c\|$ sufficiently small. The existence of such a function can be inferred by standard arguments from the regularity condition (11.4) (see e.g. [1], Ch. VII, Sec. 3). It follows now from Lemma 10.2 and its corollary that the vectors

$$(11.5) \qquad \phi(t_e,t_\nu)p_\nu, \ \nu=1,2,\ldots, \qquad \phi(t_e,t_o)c, \ c\in T_o$$

form a derived set (in the sense of Hestenes, see [10], Ch. IV, Sec.2) for $\mathcal{A}$ at $\tilde{x}(t_o)$, where $\mathcal{A}$ is the set of all points which are attainable at $t=t_e$ along admissible trajectories initiating at $t=t_o$ on M_o. Taking now the optimality of the reference trajectory into account it is easy to see that the existence of a vector $y_e \neq 0$ satisfying the relations (11.4) follows from the generalized multiplier rule (see e.g. [10], Ch. IV, Theorem 3.1).

A slight modification of the arguments used above leads to a certain refinement of Theorem 9.1. It is concerned with the question whether the requirement $p\in \Pi_t$ can be replaced by the weaker condition $p\in \Pi_t^*$; this is of interest for certain applications. The answer given in the theorem below is not a completely affirmative one, however it shows that one can allow in countably many cases that p is element of Π_t^* rather than of Π_t.

<u>Theorem 11.1.</u> Let the reference pair be optimal and let (t'_μ,p'_μ), $\mu=1,2,\ldots$ be a sequence satisfying the conditions

$$t'_\mu\in(t_o,t_e), \ t'_\mu \neq t'_\rho \quad \text{for } \mu \neq \rho, \ p'_\mu\in \Pi_{t'_\mu} .$$

Then one can find an adjoint state vector $y(\cdot)$ which is such that the statement of Theorem 9.1 holds true and that we have in addition

$$y(t'_\mu)^T p'_\mu \leq 0, \quad \mu=1,2,\ldots .$$

<u>Proof.</u> It is easy to see that one can find a sequence of pairs (t_ν,p_ν) which has all properties stated in Lemma 11.1 and besides satisfies the condition $t_\nu \neq t'_\mu$ for all ν,μ. If the vectors

$$\phi(t_e,t'_\mu)p'_\mu \ , \ \mu=1,2,\ldots$$

are added to the vectors (11.5) one obtains a derived set and the conclusion of the theorem follows as before.

PART II: INTERIOR CONTROLS. SECOND ORDER CONDITIONS.

The main results of part II are contained in Theorem 20.1, 20.2, 21.1, 21.2 respectively. The statements involve certain definitions and explanations given in Sec. 12-19. We will indicate briefly how one can acquire the information which is necessary in order to understand and apply the second order necessary conditions described in this work.

Note that the dimension of the control and state variable respectively is n and m respectively and that there is no restriction upon m. All of the subsequent considerations pertain to an arbitrary but fixed reference solution $u(\cdot),x(\cdot)$ satisfying the condition

$$u(t) \in \text{int}U, \quad t \in I ,$$

I being some open subinterval of $[t_o,t_e]$ (= the interval on which $u(\cdot),x(\cdot)$ is supposed to exist). The hypotheses of theorems mentioned above involve the notion of a linear space $\mathcal{L}(t)$. A possible way of introducing this space is via linearization of the given system around the reference solution. To this purpose consider the linear system

$$\dot{x} = A(t)x + B(t)u,$$

where

$$A(t) := f_x(t,x(t);u(t)), \quad B(t) := f_u(t,x(t);u(t)),$$

and let the sequence of matrices $B_i(t)$ (of type (n,m)) be recursively defined as follows:

$$B_o(t) := B(t), \quad B_{i+1}(t) := \dot{B}_i(t) - A(t)B_i(t) .$$

Then $\mathcal{L}(t)$ is the linear space spanned by the columns of the $B_i(t)$. Note that in case of a linear time invariant system (i.e. if $f(t,x;u) = Ax + Bu$) $\mathcal{L}(t)$ is independent of t and equals the linear space spanned by the columns of the Kalman controllability matrix. It is well known - and could easily be established with the help of our Lemma 14.2 - that the standard criterion for complete controllability (dimension $\mathcal{L}(t)$ equals n) can be extended to the linear time dependent case. As a consequence $\mathcal{L}(t)$ cannot have its maximal dimension along an optimal solution (in a suitable augmented state space). The usual way to put this into evidence is to evaluate the relation

$$H_u(t,x(t),y(t),u(t)) = 0, \quad t \in I,$$

by repeated differentiation with respect to t. (H is the Hamiltonian associated with the control problem. The above identity follows from the maximum principle and from the assumption $u(t) \in \text{int}U$). One then obtains the so called first order conditions, namely

$$y(t)^T B_i(t) = 0, \quad i=0,1,\ldots .$$

This well known fact provides the motivation for our definition of $B_i(\ldots)$ which is given in Sec. 13 (cf. in particular (13.7)). As was pointed out in Sec. 1 the B_i there will be functions of t,x and further variables $u_o,u_1,\ldots$. Therefore their derivatives with respect to the components u_o^j of u_o and also the Lie-brackets $[\ldots]$ of column vectors taken from some B_i are well defined (cf. (12.3) for the precise meaning of $[\ldots]$).This notation is widely used in Sec. 21.

The way in which the generalized Clebsch-Legendre condition is stated in Sec. 20 calls for some further explanation. The formulation is based on a parametrization of the control variable which is given in the form

$$u = u(t) + \xi v(t)$$

where ξ is scalar and $v(t)$ an arbitrary m-dimensional time-dependent vector. In case $m=1$ it makes no sense to select a $v(t)$ which is different from 1, in the case $m>1$ however this procedure enables us to present for each "direction" of the control space an individual second order test. The word "direction" seems appropriate since the above parametrization of u indeed can be viewed upon as a digression from the reference control in specified direction $v(t)$. One advantage of this approach is that we do not need the notion of "degree of singularity" which is difficult to define if $m>1$. This can be best seen in the case of free-endpoint problems. In Theorem 22.2 the "directional" version of the generalized Clebsch Legendre condition is restated for this type of problem and it then turns out that there is then also formally no difference between the cases $m=1$ and $m>1$.

The parametrization of the control variable leads to the definition of two types of quantities which are denoted by $\tilde{b}_\nu(t,x,\xi)$ and $b_\nu(t,x)$ respectively. Both are n-dimensional vectors depending upon t,x,ξ and t,x respectively. The first kind is defined in terms of the Hamiltonian formalism used already in connection with the $B_i(\ldots)$ (Sec.13). The second kind could as well be defined via the same formalism, however for the reader's convenience we prefer to write down the definition in explicit form.

A last word seems to be in order concerning equality type conditions (Sec.21). The considerations are actually carried out for $m=2$, the extension to higher dimensions is indicated at the beginning of Sec.21. The matrix B_i consists then of two column vectors which are denoted by $B_i^\nu, \nu=1,2$, and have to be regarded as functions of $t,x,u_o,u_1,\ldots$

12. Linear and Quadratic Approximations of $h^{(\nu)}$.

The study of the functions $h^{(\nu)}(t,x;\,\mathbf{U},\mathbf{V})$ which were introduced in Sec. 7 will be resumed; the notation will be the same as there. We know from Theorem 7.1 and from Lemma 7.4 (part (i)) that the relation $h^{(\nu)}(t,x;\,\mathbf{U},\mathbf{U}) = 0$ holds identically in $t,x,\mathbf{U}$ if $\nu>0$. Hence if $h^{(\nu)}$ is regarded as a function of $\mathbf{V}$ and expanded in a Taylor-series at $\mathbf{V}=\mathbf{U}$ the first two terms will be linear and second order polynomials in the components of $\mathbf{V}-\mathbf{U}$. Explicit formulae for these terms will be derived in this section.

The symbols $\mathbf{U}$ and $\mathbf{V}$ respectively represent as before two sequences of the form $\{u_o,u_1,\ldots\}$ and $\{v_o,v_1,\ldots\}$ respectively where $u_i = (u_i^{(1)},\ldots,u_i^{(m)})^T$ and $v_i = (v_i^{(1)},\ldots,v_i^{(m)})^T$ are m-dimensional vectors. The $u_i^{(\mu)}$ and $v_i^{(\mu)}$ have to be regarded as independent variables. The functions g,h etc. which appear in the sequel and which are n-dimensional vectors depend upon t,x and finitely many of the components of $\mathbf{U}$, $\mathbf{V}$ and are tacitly assumed to be of class C^∞ on the set

(12.1) $\{t,x,\,\mathbf{U},\mathbf{V} : (t,x,u_o) \in Y \text{ and } (t,x,v_o) \in Y\}$

(for the definition of Y cf. (7.1)). $f=f(t,x;u_o)$ always denotes the function which arises from the right hand side of the given differential equation (7.1) by replacing the control variable by u_o, hence f has to be regarded from now on as a function of $t,x,\mathbf{U}$. As before we frequently will omit all or some of the arguments $t,x,\mathbf{U},\mathbf{V}$ whenever this will not lead to misunderstandings. Furthermore we use a special notation to express the fact that the difference $g = g_1-g_2$ of two functions and all partial derivatives of g with respect to the components of $\mathbf{V}$ up to some order $p-1$ $(p\geq1)$ vanish on the set

(12.2) $\{t,x,\,\mathbf{U},\mathbf{V} : \mathbf{V}=\mathbf{U}\}$.

In this case we write $g = \mathcal{O}(\|\mathbf{V}-\mathbf{U}\|^p)$ or $g_1 = g_2 + \mathcal{O}(\|\mathbf{V}-\mathbf{U}\|^p)$.

If G is a matrix (depending upon $t, x, \mathcal{U}, \mathcal{V}$) we write $G = \mathcal{O}(\|\mathcal{V} - \mathcal{U}\|^p)$ if the $\mathcal{O}$-statement holds for each column of G.

<u>Lemma 12.1</u> (i) $g = \mathcal{O}(\|\mathcal{V} - \mathcal{U}\|^p)$ iff all partial derivatives of order $\leq p-1$ with respect to the components of $\mathcal{V}$ and $\mathcal{U}$ vanish on the set (12.2).

(ii) $g = \mathcal{O}(\|\mathcal{V} - \mathcal{U}\|^p)$ implies $\partial g/\partial \xi = \mathcal{O}(\|\mathcal{V} - \mathcal{U}\|^p)$ where ξ equals t or a component of x.

(iii) $g(\mathcal{U}, \mathcal{V}) = \mathcal{O}(\|\mathcal{V} - \mathcal{U}\|^p)$ implies $g(\mathcal{V}, \mathcal{U}) = \mathcal{O}(\|\mathcal{V} - \mathcal{U}\|^p)$.

(iv) $g(\mathcal{U}, \mathcal{V}) = \mathcal{O}(\|\mathcal{V} - \mathcal{U}\|^2)$ implies $g(\mathcal{U}, \mathcal{V}) - g(\mathcal{V}, \mathcal{U}) = \mathcal{O}(\|\mathcal{V} - \mathcal{U}\|^3)$.

(v) $g = \mathcal{O}(\|\mathcal{V} - \mathcal{U}\|^p)$ implies

$$\frac{\partial g}{\partial u_\rho} u_{\rho+1} + \frac{\partial g}{\partial v_\rho} v_{\rho+1} = \mathcal{O}(\|\mathcal{V} - \mathcal{U}\|^p) .$$

<u>Proof.</u> If $g = \mathcal{O}(\|\mathcal{V} - \mathcal{U}\|^p)$ then, by standard arguments based on Taylor's theorem, each component of g can locally be written as a homogeneous polynomial of degree p in the variables $v_i^{(\mu)} - u_i^{(\mu)}$, the coefficients being infinitely often differentiable functions of $\mathcal{U}, \mathcal{V}$. Using this representation one easily confirms the statements (i), (iii), (iv). (ii) follows immediately from the above explanation of the $\mathcal{O}$-relation. In order to prove (v) let us consider an arbitrary differential operator Δ acting on the components of $\mathcal{V}$. If Δ is of order $\leq p-1$ and if $g = \mathcal{O}(\|\mathcal{V} - \mathcal{U}\|^p)$ we have $(\Delta g)(\mathcal{U}, \mathcal{U}) = 0$ and hence

$$0 = \frac{\partial}{\partial u_\rho} (\Delta g)(\mathcal{U}, \mathcal{U}) = \frac{\partial}{\partial u_\rho}(\Delta g)(\mathcal{U}, \mathcal{V}) + \frac{\partial}{\partial v_\rho}(\Delta g)(\mathcal{U}, \mathcal{V}) \Big|_{\mathcal{V} \to \mathcal{U}}$$

$$= \Delta\left(\frac{\partial}{\partial u_\rho} g(\mathcal{U}, \mathcal{V}) + \frac{\partial}{\partial v_\rho} g(\mathcal{U}, \mathcal{V})\right) \Big|_{\mathcal{V} \to \mathcal{U}}$$

Since this is true for an arbitrary Δ of order $\leq p-1$ we have

$$\frac{\partial}{\partial u_\rho} g + \frac{\partial}{\partial v_\rho} g = \mathcal{O}(\| \mathcal{V} - \mathcal{U} \|^p)$$

and hence

$$\left(\frac{\partial}{\partial u_\rho} g + \frac{\partial}{\partial v_\rho} g\right)\cdot u_{\rho+1} = \mathcal{O}(\| \mathcal{V} - \mathcal{U} \|^p) .$$

Furthermore it is clear that $g = \mathcal{O}(\| \mathcal{V} - \mathcal{U} \|^p)$ implies

$$\frac{\partial g}{\partial v_\rho}\cdot(v_{\rho+1}-u_{\rho+1}) = \mathcal{O}(\| \mathcal{V} - \mathcal{U} \|^p) .$$

Adding the two last relations leads to the desired conclusion (v).

The Lie-bracket [...] which appears for the first time in the next theorem will have the usual meaning, throughout this paper, that is

(12.3) $\quad [g,h] = h_x g - g_x h .$

Here g,h are n-dimensional vectors, depending upon x and possibly other variables as $t, \mathcal{U}, \mathcal{V}$.

<u>Theorem 12.1.</u> Let $l^{(\nu)} = l^{(\nu)}(t,x; \mathcal{U}, \mathcal{V})$ be recursively defined as follows

$$l^{(0)} = 0, \quad l^{(1)} = f_u(t,x;u_0)v_0$$

$$l^{(\nu+1)} = (l^{(\nu)})_t + \sum_{\sigma=0}^{\infty}\left((\partial l^{(\nu)}/\partial u_\sigma)\cdot u_{\sigma+1} + (\partial l^{(\nu)}/\partial v_\sigma)\cdot v_{\sigma+1}\right) + [f,l^{(\nu)}],$$

where $f = f(t,x;u_0)$. Then the following statements are true. (i) $l^{(\nu)}$ depends upon $t,x,u_0,\dots,u_{\nu-1}, v_0,\dots,v_{\nu-1}$ (hence the infinite sum in the formula for $l^{(\nu+1)}$ is in fact finite) and is of class C^∞ on the set (12.1). (ii) $l^{(\nu)}$ is linear in $\mathcal{V}$ and vanishes for $\mathcal{V} = \mathcal{O}$ identically in $t,x, \mathcal{U}$. (iii) $h^{(\nu)}(t,x; \mathcal{U}, \mathcal{V}) = -l^{(\nu)}(t,x; \mathcal{U}, \mathcal{V}-\mathcal{U}) + \mathcal{O}(\| \mathcal{V} - \mathcal{U} \|^2)$ for $\nu > 0$.

<u>Proof.</u> (i) and (ii) can be inferred immediately from the definition of $l^{(\nu)}$, from (12.3) and from the fact that f does not depend upon $\mathcal{V}$. We now prove (iii) by means of induction with respect to ν. The case $\nu = 1$ is easily settled, since we have

$$h^{(1)}(t,x;\, \mathcal{U},\mathcal{V}) = -f_u(t,x;u_o)(v_o-u_o) + \mathcal{O}(\|\mathcal{V}-\mathcal{U}\|^2)$$

in view of (7.17). Let us now assume that the statement in question has been proved for all $\nu \le \mu$, where $\mu \ge 1$. We know then from part (ii) of the theorem that

(12.4) $$h^{(\nu)}(t,x;\, \mathcal{U},\mathcal{V}) = \mathcal{O}(\|\mathcal{V}-\mathcal{U}\|)$$

for $\nu \le \mu$. Using this relation and hypothesis of induction we see from Lemma 3.1 that the composite function

$$(\mathcal{D}^{\mu}_{x} f)(t,x_o,\dots,x_\mu;v_o)\Big|_{x_\lambda \to h^{(\lambda)}}$$

is equal to

$$f_x(t,x;v_o)h^{(\mu)}(t,x;\, \mathcal{U},\mathcal{V}) + \mathcal{O}(\|\mathcal{V}-\mathcal{U}\|^2) =$$
$$=-f_x(t,x;u_o)l^{(\mu)}(t,x;\, \mathcal{U},\mathcal{V}-\mathcal{U}) + \mathcal{O}(\|\mathcal{V}-\mathcal{U}\|^2)\,.$$

Comparison of the recursive systems for $l^{(\nu)}$ and for $h^{(\nu)}$ (cf. (7.16))and the usage of the rules (ii), (v) of Lemma 12.1 leads then to the desired relation for $\nu = \mu+1$.

Next we introduce for every $\nu = 1,2,\dots$ a further (vector-valued) function of three "variables" $\mathcal{U},\mathcal{V},\mathcal{W}$. $\mathcal{W}$ again is an infinite sequence .This function is a linear homogeneous form in $\mathcal{W}$ and hence will be written in the form $L^{(\nu)}\cdot\mathcal{W}$, where $L^{(\nu)}$ is a matrix of n rows and countably many columns. With finitely many exceptions all columns of $L^{(\nu)}$ are zero, identically in t,x,

<u>Definition 12.1.</u> Let $\mathcal{W} = \{w_o,w_1,\dots\}$ be a further sequence of independent variables, each $w_i = (w_i^{(1)},\dots,w_i^{(m)})^T$ being again a m-dimensional vector. By $L^{(\nu)}(t,x;\, \mathcal{U},\mathcal{V})\cdot\mathcal{W}$ we denote the n-dimensional column vector given by

(12.5) $$\sum_{j=0}^{\infty} (\partial l^{(\nu)}(t,x;\, \mathcal{U},\mathcal{V})/\partial u_j)\cdot w_j\,.$$

Note that this sum is actually finite, since $l^{(\nu)}$ depends upon u_j with $j<\nu$ only. At later occasions we will make use of the following relation (the arguments t,x are omitted)

$$(12.6) \quad l^{(\nu)}(\mathbf{V}, \mathbf{U}-\mathbf{V}) + l^{(\nu)}(\mathbf{U}, \mathbf{V}-\mathbf{U}) = -L^{(\nu)}(\mathbf{U}, \mathbf{V}-\mathbf{U})\cdot(\mathbf{V}-\mathbf{U}) + \mathcal{O}(\|\mathbf{v}-\mathbf{u}\|^3)$$

The proof is straightforward. Since $l^{(\nu)}(\mathbf{U}, \mathbf{V})$ is linear in $\mathbf{V}$ it can be written as a linear combination of n-dimensional vectors depending upon $\mathbf{U}$ (and t,x) with the components of the v_i as coefficients:

$$l^{(\nu)}(\mathbf{U}, \mathbf{V}) = \sum_{i,\mu} k_{i,\mu}(\mathbf{U}) v_i^{(\mu)} .$$

Combining this with (12.5) we arrive at the relations

$$L^{(\nu)}(\mathbf{U},\mathbf{V})\cdot\mathbf{W} = \sum_{i,\mu}\Big(\sum_j (\partial k_{i,\mu}(\mathbf{U})/\partial u_j) w_j\Big) v_i^{(\mu)} ,$$

$$l^{(\nu)}(\mathbf{V}, \mathbf{U}-\mathbf{V}) + l^{(\nu)}(\mathbf{U}, \mathbf{V}-\mathbf{U}) = \sum_{i,\mu} (k_{i,\mu}(\mathbf{V}) - k_{i,\mu}(\mathbf{U}))(u_i^{(\mu)} - v_i^{(\mu)})$$

$$= -\sum_{i,\mu}\Big(\sum_j (\partial k_{i,\mu}(\mathbf{U})/\partial u_j)(v_j - u_j) + \mathcal{O}(\|\mathbf{V}-\mathbf{U}\|^2)\Big)(v_i^{(\mu)} - u_i^{(\mu)}) ,$$

from which (12.6) can be obtained immediately. In passing we state a further relation which follows from the last formula, namely

$$(12.7) \quad l^{(\nu)}(\mathbf{V}, \mathbf{U}-\mathbf{V}) + l^{(\nu)}(\mathbf{U}, \mathbf{V}-\mathbf{U}) = \mathcal{O}(\|\mathbf{V}-\mathbf{U}\|^2) .$$

<u>Theorem 12.2</u> We have, for $\nu = 1,2,\ldots$

$$h^{(\nu)}(t,x;\mathbf{U},\mathbf{V}) + l^{(\nu)}(t,x;\mathbf{U},\mathbf{V}-\mathbf{U}) = -\tfrac{1}{2}L^{(\nu)}(t,x;\mathbf{U},\mathbf{V}-\mathbf{U})\cdot(\mathbf{V}-\mathbf{U}) +$$
$$+ \frac{1}{2}\sum_{\sigma+\rho=\nu}\binom{\nu}{\sigma} l_x^{(\rho)}(t,x;\mathbf{U},\mathbf{V}-\mathbf{U})\cdot l^{(\sigma)}(t,x;\mathbf{U},\mathbf{V}-\mathbf{U}) + \mathcal{O}(\|\mathbf{V}-\mathbf{U}\|^3)$$

($l_x^{(\rho)}$ = Jacobian matrix of $l^{(\rho)}$ with respect to x).

<u>Proof.</u> The proof will be based on a closer study of certain of the functions $K_{\underline{\nu}}$ introduced in the beginning of Sec. 8. We take the specializations

$$N = 2, \quad \underline{\nu} = (\sigma,\rho),\ \sigma>0, \rho>0, \quad \mathbf{U}_0 = \mathbf{U}, \quad \mathbf{U}_1 = \mathbf{V}, \quad \mathbf{U}_2 = \mathbf{U} .$$

and apply Theorem 7.1 with $D' = D_1^{\sigma}, D'' = D_2^{\rho}$ (for the meaning of D_i cf. Theorem 5.1). This gives us the formula

$$(12.8)\quad K_{(\sigma,\rho)} = \mathcal{D}_x^{\sigma} h^{(\rho)}(t,x_0,\dots,x_\rho;\ \mathbf{V},\mathbf{U})\Big|_{x_\lambda \to h^{(\lambda)}(t,x;\ \mathbf{U},\mathbf{V})}$$

and consequently, in view of Lemma 3.1, and (12.4) the relation

$$(12.9)\quad K_{(\sigma,\rho)} = h_x^{(\rho)}(t,x;\ \mathbf{V},\mathbf{U})h^{(\sigma)}(t,x;\ \mathbf{U},\mathbf{V}) + \mathcal{O}(\|\mathbf{V}-\mathbf{U}\|^3).$$

Note that all partial derivatives of h with respect to x are of order $\mathcal{O}(\|\mathbf{V}-\mathbf{U}\|)$ (cf. Lemma 12.1, part (ii)). Using part (iii) of the same lemma and (12.7) one infers from Theorem 12.1, part (ii) and (iii) that

$$h^{(\rho)}(t,x;\ \mathbf{V},\mathbf{U}) = l^{(\rho)}(t,x;\ \mathbf{U},\mathbf{V}-\mathbf{U}) + \mathcal{O}(\|\mathbf{V}-\mathbf{U}\|^2).$$

Applying Theorem 12.1 and the rules given in Lemma 12.1 once more (12.9) can be changed to

$$K_{(\sigma,\rho)}(t,x;\ \mathbf{U},\mathbf{V},\mathbf{U}) = -l_x^{(\rho)}(t,x;\ \mathbf{U},\mathbf{V}-\mathbf{U})l^{(\sigma)}(t,x;\mathbf{U},\mathbf{V}-\mathbf{U}) + \mathcal{O}(\|\mathbf{V}-\mathbf{U}\|^3).$$

Note that this is true under the proviso $\sigma>0$, $\rho>0$. We now remind the reader on two basic relations obtained in Sec. 7, namely

$$K_{(\sigma,\rho)}(t,x;\ \mathbf{U},\mathbf{V},\mathbf{U}) = \begin{cases} h^{(\nu)}(t,x;\ \mathbf{U},\mathbf{V}) & \text{if } \sigma=\nu, \rho=0 \\ h^{(\nu)}(t,x;\ \mathbf{V},\mathbf{U}) & \text{if } \sigma=0, \rho=\nu . \end{cases}$$

(Theorem 7.1) and

$$\sum_{\rho+\sigma=\nu} \frac{1}{\sigma!\rho!} K_{(\sigma,\rho)}(t,x;\ \mathbf{U},\mathbf{V},\mathbf{U}) = 0 \quad \text{if} \quad \nu > 0.$$

(cf. Lemma 7.4). Combining the last three relations we finally arrive at this result (t,x again are omitted)

$$(12.9)\quad h^{(\nu)}(\mathbf{U},\mathbf{V}) + h^{(\nu)}(\mathbf{V},\mathbf{U}) = \sum_{\sigma+\rho=\nu} \binom{\nu}{\sigma} l_x^{(\rho)}(\mathbf{U},\mathbf{V}-\mathbf{U})l^{(\sigma)}(\mathbf{U},\mathbf{V}-\mathbf{U}) + \mathcal{O}(\|\mathbf{V}-\mathbf{U}\|^3).$$

Note that the summation on the right hand side actually extends over the $\sigma>0$, $\rho>0$ only, since by definition $l^{(0)}=0$.

Next we observe that

$$h^{(\nu)}(\mathbf{U},\mathbf{V}) + l^{(\nu)}(\mathbf{U},\mathbf{V}-\mathbf{U}) - h^{(\nu)}(\mathbf{V},\mathbf{U}) - l^{(\nu)}(\mathbf{V},\mathbf{U}-\mathbf{V}) = \\ = \mathcal{O}(\|\mathbf{V}-\mathbf{U}\|^3) .$$

This is a consequence of Theorem 12.1, part (iii), and the rule (iv) of Lemma 12.1. Using (12.6) the last relation can be brought into this form

$$h^{(\nu)}(\mathbf{u},\mathbf{v})+2l^{(\nu)}(\mathbf{u},\mathbf{v}-\mathbf{u})-h^{(\nu)}(\mathbf{v},\mathbf{u})+L^{(\nu)}(\mathbf{u},\mathbf{v}-\mathbf{u})\cdot(\mathbf{v}-\mathbf{u}) = = \mathcal{O}(\|\mathbf{v}-\mathbf{u}\|^3) .$$

Adding this to (12.9) yields the statement of the theorem.

13. The operator Γ .

We associate with the control system (7.1) a linear operator $\Gamma=\Gamma_f$ with acts on all (vector-valued) functions $g=g(t,x,\mathbf{u})$. Without mentioning this explicitely we always assume that these functions depend upon t,x and finitely many elements of the sequence $\mathbf{u} = \{u_0,u_1,\dots\}$ and are of class C^∞ whenever $(t,x,u_0)\in Y$. The definition of Γ is then as follows

(13.1) $\Gamma: g\mapsto\Gamma(g)=\partial g/dt+\sum_{i=0}^{\infty}(\partial g/\partial u_i)\cdot u_{i+1}+[f,g]$

where f and the Lie-bracket $[\;\;]$ have the standard meaning explained in the previous section (cf. (12.3) and what was said in connection with (12.1)). Another way of defining Γ is via the relation

(13.2) $\frac{d}{dt}\left(y^T g(t,x,\mathbf{u})\right) = y^T\cdot(\Gamma(g))(t,x,\mathbf{u})$

which has to be understood in this way: If the scalar function $y^T g(t,x,\mathbf{u})$ is formally differentiated with respect to t and if the derivatives $\dot{x},\dot{y},\dot{u}_i$ are expressed in terms of $t,x,y,\mathbf{u}$ according to the rules

(13.3) $\dot{x} = f(t,x;u_0)$, $\dot{y}=-f_x(t,x;u_0)^T y$, $\dot{u}_i=u_{i+1}$, $i=0,1,\dots$

then one obtains a scalar function which can be represented in the form $y^T\cdot\Gamma(g)$. The following lemma provides us with some basic information about the operator Γ .

Lemma 13.1. (i) $g=0$ identically in t,x,u_0 on the set $\{t,x,\mathbf{u} : u_i=0 \text{ for } i>0\}$ implies $\Gamma(g) = 0$ on the same set.

(ii) If g depends upon $u_o,\dots,u_i$ only, then $\Gamma(g)$ depends upon $u_o,\dots,u_{i+1}$ only. (iii) $\Gamma(\alpha g) = \alpha\Gamma(g)+\beta g$ if $\alpha=\alpha(t,x,\mathcal{U})$ is scalar and

$$\beta=\partial\alpha/\partial t + \sum_{i=0}^{\infty} (\partial\alpha/\partial u_i)^T\cdot u_{i+1} + (\partial\alpha/\partial x)^T\cdot f$$

($\partial\alpha/\partial u_i, \partial\alpha/\partial x$ etc. is the gradient of α with respect to u_i,x etc). (iv) The following relations hold for $\mu=1,\dots,m$:

$$(13.4)\quad \begin{aligned} \partial\Gamma(g)/\partial u_j^{(\mu)} &= \Gamma(\partial g/\partial u_j^{(\mu)}) + \partial g/\partial u_{j-1}^{(\mu)} \quad \text{if } j>0 \ ,\\ \partial\Gamma(g)/\partial u_o^{(\mu)} &= \Gamma(\partial g/\partial u_o^{(\mu)})+[\partial f/\partial u_o^{(\mu)},g] \ . \end{aligned}$$

(v) Let $u(t)$ be a function of class C^∞ on some open interval I, $\mathcal{U}(t)$ the corresponding sequence of the form (7.5) and $x(t)$ a solution of $\dot{x} = f(t,x;u(t))$. If the relation $g(t,x(t),\mathcal{U}(t))=0$ holds on I then we have also $(\Gamma(g))(t,x(t),\mathcal{U}(t))=0$ for all $t\in I$.

Proof. (i)-(iv) follow immediately from the definition of Γ (cf. (13.1)) and from the identity

$$(13.5)\quad [f,\alpha g] = \alpha[f,g] + ((\partial\alpha/\partial x)^T\cdot f)g \ .$$

(v) is a consequence of (13.2) and (13.3). Indeed, if we have $g(t,x(t),\mathcal{U}(t))=0$ identically in t, then $y(t)^T\cdot(\Gamma(g))(t,x(t),\mathcal{U}(t))$ must vanish for an a r b i t r a r y solution $y(\cdot)$ of the differential eq. $\dot{y} = -f_x(t,x(t),u(t))^T y$. This of course can only be true if $(\Gamma(g))(t,x(t),\mathcal{U}(t)=0$.

Corollary 1. Given a set $\mathcal{B}$ of n-dimensional column vectors $b=b(t,x,\mathcal{U})$ which is invariant under the operator Γ . Let $I,u(t)$, $\mathcal{U}(t)$ and $x(t)$ have the same meaning as in part (v) of Lemma 13.1 and let $\mathcal{B}(t)$ be the linear space spanned by the vectors $b(t,x(t),\mathcal{U}(t))$. Assume that the dimension of $\mathcal{B}(t)$ is independent from t .

Claim: $g(t,x(t),\, u(t)) \in \mathcal{L}(t)$ for all $t \in I$ implies $(\Gamma(g))(t,x(t),\, u(t)) \in \mathcal{L}(t)$ for all $t \in I$.

Proof. For the purpose of the proof the interval I can be replaced by a sufficiently small neighborhood $\mathcal{N}$ of a fixed $t_0 \in I$. Making use of the hypothesis concerning the dimension of $\mathcal{L}(t)$ and of continuity arguments we may assume that there exists a fixed set of elements of $\mathcal{L}(t)$, say $b_1(t,x,u),\dots,b_s(t,x,u)$ such that the $b_i(t,x(t),\, u(t))$ form a basis of $\mathcal{L}(t)$ for every $t \in \mathcal{N}$. Hence we have a representation of $g(t,x(t),\, u(t))$ as a linear combination $\sum_{i=1}^{s} \lambda_i(t) b_i(t,x(t),\, u(t))$ where the λ_i are sufficiently smooth scalar functions of t. In other words

$$\hat{g}(t,x,u) := g(t,x,u) - \sum_{i=1}^{s} \lambda_i(t) b_i(t,x,u)$$

vanishes for $x=x(t)$, $u = u(t)$. From part (v) of the preceding lemma one infers then that $\Gamma(g)$ coincides with $\sum_{i=1}^{s} \Gamma(\lambda_i b_i)$ along $(x,u) = (x(t),\, u(t))$. One the other hand if Γ is applied to a scalar multiple $\lambda(t)b(t,x,u)$ of some vector b one obtains in view of part (iii) of Lemma 13.1

$$\lambda(t)(\Gamma(b))(t,x,u) + \dot{\lambda}(t)b(t,x,u) .$$

If b belongs to $\mathcal{L}$, then the same is true of $\Gamma(b)$ and hence the above expression becomes an element of $\mathcal{L}(t)$ if (x,u) is replaced by $(x(t),\, u(t))$. Hence we have also $\sum_{i=1}^{s} \Gamma(\lambda_i b_i) \in \mathcal{L}(t)$ for $(x,u) = (x(t),\, u(t))$ and the corollary is proved.

The ρ-th iterate of the operator Γ is denoted by Γ^{ρ}. Application of this operator involves repeated application of the operator $g \to [f,g]$ (forming the Lie-bracket with the given function f). We adopt the standard symbols to denote the images of g under the iterates of the second operator, namely

(13.6) $\quad \mathrm{ad}^0(f)g = g, \quad \mathrm{ad}^{\rho}(f)g = [f,\, \mathrm{ad}^{\rho-1}(f)g], \quad \rho=1,2,\dots$

(cf. [3]). The next corollary is a straightforward application of the preceding considerations.

Corollary 2. Assume that $f = f(x;u_o)$ and $g = g(x, \mathfrak{u})$ do not depend explicitly upon t. Then $\Gamma(g)$ does also not depend upon t. Furthermore the function

$$\Gamma^{\rho}(g) - ad^{\rho}(f)g$$

vanishes on the set $\{t,x, \mathfrak{u} : u_i = 0 \quad \text{for} \quad i>0\}$, $\rho = 1,2,\ldots$

Proof. The first statement and the second one for $\rho=1$ follow immediately from the definition (13.1) of Γ. The proof is easily completed using induction with respect to ρ and part (i) of Lemma 13.1 .

We now present the definition of certain quantities which are associated with a given control system (7.1) and from which all information pertaining to first and second order conditions can be derived. These quantities are matrices of n rows and m colum s and will be denoted by $B_\nu = B_\nu(t,x, \mathfrak{u})$. The definiton is recursively and runs as follows

$$\text{(13.7)} \qquad \begin{aligned} B_o(t,x,\mathfrak{u}) &= (\partial f/\partial u)(t,x;u_o) \\ B_{\nu+1}(t,x,\mathfrak{u}) &= (\Gamma(B_\nu))(t,x, \mathfrak{u}), \ \nu=0,1,\ldots , \end{aligned}$$

(we agree to understand the application of Γ to a matrix as the application of Γ to each of its columns). As an immediate consequence of this definition we have this first statement on the B_ν:

(13.8) The elements of the matrix B_ν are polynomials in the components of $u_1,\ldots,u_\nu$, if $\nu\geq 1$. The coefficients are partial derivatives of the components of $f(t,x;u)$ evaluated at $u=u_o$.

An alternative way of defining B_ν can be based on the identity (13.2). Indeed, using induction with respect to ν , one easily infers from (13.7) that these relations hold for $\nu=0,1,\ldots$

$$(13.9)\qquad \frac{d^\nu}{dt^\nu}(\partial H/\partial u_0) = y^T B_\nu(t,x,\mathbf{U}), \text{ where } \quad H = H(t,x,y,\mathbf{U}) = y^T\cdot f(t,x;u_o)$$

and where d/dt means differentiation with respect to (13.3). There is still another possibility of introducing the B_ν, namely in terms of the functions $l^{(\nu)}(t,x;\mathbf{U},\mathbf{V})$ which were defined in the previous section.

Theorem 13.1 We have, for $\nu=1,2,\ldots,$

$$l^{(\nu)}(t,x;\mathbf{U},\mathbf{V}) = \sum_{j=0}^{\nu-1}\binom{\nu-1}{j}B_{\nu-1-j}(t,x,\mathbf{U})\cdot v_j$$

(note that v_j is the j-th element in the sequence $\mathbf{V}$).

Proof. By induction with respect to ν, using the definition of the $l^{(\nu)}$ as given in Theorem 12.1. The case $\nu=1$ is settled by inspection. Comparing (13.1) and the recursive definition of $l^{(\nu+1)}$ one sees that the latter can be written in this form

$$l^{(\nu+1)} = \Gamma(l^{(\nu)}) + \sum_{\sigma=0}^{\infty}(\partial l^{(\nu)}/\partial v_\sigma)\cdot v_{\sigma+1} \quad .$$

Hence we obtain from hypothesis of induction

$$l^{(\nu+1)}(t,x;\mathbf{U},\mathbf{V}) = \sum_{j=0}^{\nu-1}\binom{\nu-1}{j}(B_{\nu-j}\cdot v_j + B_{\nu-1-j}\cdot v_{j+1})$$

$$= \sum_{j=0}^{\nu}\binom{\nu}{j}B_{\nu-j}\cdot v_j \;.$$

Thereby the theorem is proved.

We conclude this section by presenting an important relation involving the B_ν. By Γ^ρ we denote as before the ρ-th iterate of the mapping Γ.

Lemma 13.2 . We have

$$\partial B_{i+j}/\partial u_j^{(\mu)} = \sum_{\rho=1}^{i}\binom{j+\rho-1}{\rho}\Gamma^\rho(\partial B_{i-\rho}/\partial u_o^{(\mu)}) + \partial B_i/\partial u_o^{(\mu)}$$

for all $i,j \geq 0$ and $\mu = 1,\ldots,m$.

Proof. We omit for shortness the superscript μ . The formula is certainly true for $j=0$ if we adopt the usual convention $\binom{\rho-1}{\rho}=0$. So we may assume from now on that $j>0$. To begin with let us consider the case $i=0$. It follows from the first of the relations (13.4) and for $g=B_{j-1}$ that

$$\partial B_j/\partial u_j = \Gamma(\partial B_{j-1}/\partial u_j) + \partial B_{j-1}/\partial u_{j-1} \ .$$

But B_{j-1} does not depend upon u_j (cf. (13.8)). Hence we have, for every $j>0$

$$\partial B_j/\partial u_j = \partial B_{j-1}/\partial u_{j-1} = \dots = \partial B_o/\partial u_o ,$$

and this is the statement of the lemma in case $i=0$, $j>0$.

The remaining cases can now be treated by double induction, first with respect to j, then - for fixed j - with respect to i. We may assume that $i>0$, $j>0$, because of what we have established already. Using the relations (13.4) once more - with $g = B_{i+j-1}$ - we obtain

$$\partial B_{i+j}/\partial u_j = \Gamma(\partial B_{i+j-1}/\partial u_j) + \partial B_{i+j-1}/\partial u_{j-1} \ .$$

From hypothesis of induction one can now derive representations for the expressions on the right hand side of the above formula, namely

$$\partial B_{i-1+j}/\partial u_j = \sum_{\rho=1}^{i-1} \binom{j+\rho-1}{\rho}\Gamma^\rho(\partial B_{i-1-\rho}/\partial u_o) + \partial B_{i-1}/\partial u_o ,$$

$$\partial B_{i+j-1}/\partial u_{j-1} = \sum_{\rho=1}^{i} \binom{j+\rho-2}{\rho}\Gamma^\rho(\partial B_{i-\rho}/\partial u_o) + \partial B_i/\partial u_o \ .$$

Applying Γ to the first and then adding the second line finally leads to this formula

$$\begin{aligned}
\partial B_{i+j}/\partial u_j &= \sum_{\rho=2}^{i} \binom{j+\rho-2}{\rho-1}\Gamma^\rho(\partial B_{i-\rho}/\partial u_o) + \Gamma(\partial B_{i-1}/\partial u_o) \\
&+ \sum_{\rho=1}^{i} \binom{j+\rho-2}{\rho}\Gamma^\rho(\partial B_{i-\rho}/\partial u_o) + \partial B_i/\partial u_o \\
&= \sum_{\rho=1}^{i} \binom{j+\rho-1}{\rho}\Gamma^\rho(\partial B_{i-\rho}/\partial u_o) + \partial B_i/\partial u_o
\end{aligned}$$

which is the desired result. In passing we note that it can be written in the form

$$(13.10) \qquad \partial B_{i+j}/\partial u_j^{(\mu)} = \sum_{\rho=0}^{i} \binom{j+\rho-1}{\rho} \Gamma^{\rho}(\partial B_{i-\rho}/\partial u_o^{(\mu)})$$

provided $j>0$.

We conclude this section by stating a lemma which later will be needed for technical purposes. Let us consider along with the given system the augmented system (8.12), that is the system

(13.11) $\dot{x}^* = f^*(t,x^*,u)$, where $x^*=(x,x^{n+1})^T$, $f^* = (f,c)^T$,

c being a constant. The operator $\Gamma^*=\Gamma_{f^*}$ which is associated with (13.11) acts on all (n+1)-dimensional vectors $g^*=g^*(t,x^*,\boldsymbol{u})$.

Lemma 13.3 If g^* has a vanishing (n+1)-th component, so does $\Gamma^*(g^*)$. If g^* is of the form $g^*(=(g,0)^T$, where $g=g(t,x,\boldsymbol{u})$ does not depend upon x^{n+1} , then $\Gamma^*(g^*) = (\Gamma(g),0)^T$.

Proof. Follows from definition 13.1. - From Theorem 13.1 we conclude that the matrices B^*_ν associated with the augmented system (13.11) have zeros in the last row and otherwise coincide with B_ν . Hence it is also clear that the quantities which play the role of $l^{(\nu)}$ and $L^{(\nu)}\cdot\boldsymbol{w}$ for the system (13.11) are simply the (n+1)-dimensional vectors $(l^{(\nu)},0)^T$ and $(L^{(\nu)}\cdot\boldsymbol{w},0)^T$ respectively.

14. Linear and Quadratic Approximation of $K_{\underline{\nu}}$.

We resume the study of the functions

$$K_{\underline{\nu}} = K_{\underline{\nu}}(t,x;\ \boldsymbol{u}_o,\ \boldsymbol{u}_1,\ldots,\ \boldsymbol{u}_N)$$

which were introduced in Section 8. The integer N has to be regarded as fixed in the following. $\underline{\nu}$ denotes a N-tupel $(\nu_1,\ldots,\nu_N)$ of non-negative integers. As we have remarked earlier the functions $K_{\underline{\nu}}$ vanish identically in $t,x,\ \boldsymbol{u}_o$ if $\boldsymbol{u}_i$ is replaced by $\boldsymbol{u}_o$

for $i=1,\dots,N$ (cf. Lemma 7.4). Hence the Taylor-expansion of $K_{\underline{\nu}}$ - regarded as a function of $\mathfrak{U}_1,\dots,\mathfrak{U}_N$ - at $\mathfrak{U}_i = \mathfrak{U}_o$, $i=1,\dots,N$, begins with linear terms. We want to give in this section an explicit formula for the first and second order Taylor-polynomials $K^{(1)}_{\underline{\nu}}$ and $K^{(2)}_{\underline{\nu}}$ respectively. The precise definition of $K^{(\lambda)}_{\underline{\nu}} = K^{(\lambda)}_{\underline{\nu}}(t,x;\mathfrak{U}_o,\dots,\mathfrak{U}_N)$ which will be used in the sequel runs as follows: $K^{(o)}_{\underline{\nu}}=0$, $K^{(\lambda)}_{\underline{\nu}}$ is a polynomial of degree λ in $\mathfrak{U}_i - \mathfrak{U}_o$, $i=1,\dots,N$, the coefficients being C^∞ functions of $t,x,\mathfrak{U}_o$ if $\lambda>0$. The difference $K_{\underline{\nu}} - K^{(\lambda)}_{\underline{\nu}}$ and all its partial derivatives with respect to t,x is of order $\mathcal{O}\left(\sum_{i=1}^{N}\|\mathfrak{U}_i-\mathfrak{U}_o\|^{\lambda+1}\right)$.

As an immediate consequence of this definition we state the

<u>Lemma 14.1</u> If more than λ components of $\underline{\nu}=(\nu_1,\dots,\nu_N)$ are positiv, then $K^{(\lambda)}_{\underline{\nu}}=0$, i.e. $K_{\underline{\nu}} = \mathcal{O}\left(\sum_{i=1}^{N}\|\mathfrak{U}_i-\mathfrak{U}_o\|^{\lambda+1}\right)$.

<u>Proof.</u> The statement is certainly true for $\lambda = 0$. So we proceed by induction with respect to λ. If $\lambda>1$ we infer from Theorem 7.1 that $K_{\underline{\nu}}$ admits a representation of the form

$$(14.1)\quad \left(\mathcal{D}^{\mu}_{x}K_{\underline{\nu}'}\right)(t,x_o,x_1,\dots,x_\mu;\mathfrak{U}_o,\mathfrak{U}_1,\dots,\mathfrak{U}_N)$$
$$\text{with}\quad x_\rho \to h^{(\rho)}(t,x;\mathfrak{U}_{i-1},\mathfrak{U}_i)\ ,\ \rho=0,\dots,\mu,$$

where the multiindex $\underline{\nu}'$ differs from $\underline{\nu}$ by just one component. According to hypothesis of induction $K_{\underline{\nu}'}$ and all partial derivatives with respect to t and x are of order $\mathcal{O}\left(\sum_{i=1}^{N}\|\mathfrak{U}_i-\mathfrak{U}_o\|^{\lambda}\right)$. Further $h^{(\rho)}$ vanishes identically in $t,x,\mathfrak{U}_o$ if $\mathfrak{U}_i=\mathfrak{U}_o$, $i=1,\dots,N$. It follows then from Lemma 3.1 that the expression (14.1) is of order $\mathcal{O}\left(\sum_{i=1}^{N}\|\mathfrak{U}_i-\mathfrak{U}_o\|^{\lambda+1}\right)$ and the lemma is proved.

Theorem 14.1 (i) If $\underline{\nu}=(0,\dots,0,\nu,0,\dots,0)$ where ν appears at place i then $K^{(1)}_{\underline{\nu}} = 1^{(\nu)}(t,x;\ \boldsymbol{U}_o,\ \boldsymbol{U}_{i-1}-\boldsymbol{U}_i)$ and

$$K^{(2)}_{\underline{\nu}} = K^{(1)}_{\underline{\nu}} + \tfrac{1}{2} L^{(\nu)}(t,x;\ \boldsymbol{U}_o,\ \boldsymbol{U}_{i-1}-\boldsymbol{U}_i)\cdot(\boldsymbol{U}_{i-1}+\boldsymbol{U}_i-2\boldsymbol{U}_o)$$

$$+ \tfrac{1}{2}\sum_{\sigma+\rho=\nu} \binom{\nu}{\sigma} 1^{(\rho)}_x(t,x;\ \boldsymbol{U}_o,\ \boldsymbol{U}_{i-1}-\boldsymbol{U}_i)\cdot 1^{(\sigma)}(t,x;\ \boldsymbol{U}_o,\ \boldsymbol{U}_{i-1}-\boldsymbol{U}_i).$$

(ii) If $\underline{\nu} = (0,\dots,0,\sigma,0,\dots,0,\rho,0,\dots,0)$ where $\sigma>0$ appears at place i and $\rho>0$ at place $j>i$ then

$$K^{(2)}_{\underline{\nu}} = 1^{(\rho)}_x(t,x;\ \boldsymbol{U}_o,\ \boldsymbol{U}_{j-1}-\boldsymbol{U}_j)\cdot 1^{(\sigma)}(t,x;\ \boldsymbol{U}_o,\ \boldsymbol{U}_{i-1}-\boldsymbol{U}_i)\ .$$

Proof. We omit the argument t,x occasionally throughout the remaining portion of this section. The first half of (i) follows simply from the fact that if $\underline{\nu}$ is of the form as specified above then

(14.2) $$K_{\underline{\nu}}=h^{(\nu)}(\boldsymbol{U}_{i-1},\ \boldsymbol{U}_i) = 1^{(\nu)}(\boldsymbol{U}_o,\ \boldsymbol{U}_{i-1}-\boldsymbol{U}_i)+\mathcal{O}\Big(\sum_{i=1}^{N}\|\boldsymbol{U}_i-\boldsymbol{U}_o\|^2\Big)$$

(cf. Theorem 7.1 and Theorem 12.1, part (iii)).

The second half follows from the same representation of $K_{\underline{\nu}}$ and from Theorem 12.2. Note that the following relations hold up to terms of order $\mathcal{O}\big(\sum_{i=1}^{N}\|\boldsymbol{U}_i-\boldsymbol{U}_o\|^3\big)$:

$$1^{(\nu)}(\boldsymbol{U}_{i-1},\boldsymbol{U}_{i-1}-\boldsymbol{U}_i) = 1^{(\nu)}(\boldsymbol{U}_o,\boldsymbol{U}_{i-1}-\boldsymbol{U}_i) + $$
$$+ L^{(\nu)}(\boldsymbol{U}_o,\boldsymbol{U}_{i-1}-\boldsymbol{U}_i)\cdot(\boldsymbol{U}_{i-1}-\boldsymbol{U}_o)$$

(cf. Definition 12.1), and

$$L^{(\nu)}(\boldsymbol{U}_{i-1},\boldsymbol{U}_i-\boldsymbol{U}_{i-1})\cdot(\boldsymbol{U}_i-\boldsymbol{U}_{i-1}) =$$

$$= L^{(\nu)}(\boldsymbol{U}_o,\boldsymbol{U}_{i-1}-\boldsymbol{U}_i)\cdot(\boldsymbol{U}_{i-1}-\boldsymbol{U}_i).$$

In order to prove (ii) we refer to Theorem 7.1. Taking $D'=D_i^{\sigma}$, $D''=D_j^{\rho}$ we obtain the representation

$$K_{\underline{\nu}}=(\mathcal{D}_x^{\sigma}h^{(\rho)})(t,x_0,\ldots,x_\sigma;\ \mathbf{u}_{j-1},\ \mathbf{u}_j)\Big|_{x_\lambda\to h^{(\lambda)}(t,x;\ \mathbf{u}_{i-1},\ \mathbf{u}_i)}\quad .$$

The desired result follows then from (14.2) by the same type of argument which was used in connection with (12.8), (12.9).

We introduce the symbol $C^{(\lambda)}(t,x,z;\ \mathbf{u}_0,\ldots,\ \mathbf{u}_N)$ in order to denote the formal power series

$$x+\sum_{|\underline{\nu}|>0}\frac{1}{\underline{\nu}!}K_{\underline{\nu}}^{(\lambda)}(t,x;\ \mathbf{u}_0,\ldots,\ \mathbf{u}_N)z^{\underline{\nu}}$$

From the definition of $K_{\underline{\nu}}^{(\lambda)}$ it is clear that the difference between the finite partial sums $\sum\limits_{0<|\underline{\nu}|<\rho}$ of $C^{(\lambda)}$ and those of C (cf. (8.2)) is of order $\mathcal{O}(\sum\limits_{i=1}^{N}\|\mathbf{u}_i-\ \mathbf{u}_0\|^{\lambda+1})$. It is not difficult, in view of Lemma 14.1 and Theorem 14.1, to set up the explicit formula for $C^{(1)}$ and $C^{(2)}$ (the argument t,x again is partly omitted).

$$C^{(1)}=x+\sum_{\nu=1}^{\infty}\frac{1}{\nu!}\Big(\sum_{i=1}^{N}1^{(\nu)}(\ \mathbf{u}_0,\ \mathbf{u}_{i-1}-\ \mathbf{u}_i)z_i^{\nu}\Big)$$

(14.3) $$C^{(2)}=C^{(1)}+\frac{1}{2}\sum_{\nu=1}^{\infty}\frac{1}{\nu!}\Big(\sum_{i=1}^{N}L^{(\nu)}(\ \mathbf{u}_0,\ \mathbf{u}_{i-1}-\mathbf{u}_i)\cdot(\mathbf{u}_{i-1}+\mathbf{u}_i-2\,\mathbf{u}_0)z_i^{\nu}\Big)+$$

$$+\check{C}^{(2)},$$

where

$$\check{C}^{(2)}=\frac{1}{2}\sum_{\nu=1}^{\infty}\ \sum_{\sigma+\rho=\nu}\ \sum_{i=1}^{N}\mathcal{L}_i^{\rho}\cdot 1_i^{\sigma}\ +$$

$$+\sum_{\sigma>0}\ \sum_{\rho>0}\ \sum_{1\le i<j\le N}\mathcal{L}_j^{\rho}\cdot 1_i^{\sigma}\quad .$$

Here we have used the abbreviations for the moment

$$(14.4)\qquad \begin{aligned} l_i^\nu &:= \frac{1}{\nu!}\, l^{(\nu)}(t,x;U_o,\, U_{i-1}-U_i)z_i^\nu \quad, \\ \mathcal{L}_i^\nu &:= \frac{1}{\nu!}\, l_x^{(\nu)}(t,x;U_o,\, U_{i-1}-U_i)z_i^\nu \quad. \end{aligned}$$

Note that l_i^ν is a n-dimensional column vector, $\mathcal{L}_i^\nu$ a $n\times n$ matrix. Since $l^{(o)} = 0$ (cf. Theorem 12.1) the expression for $\check{C}^{(2)}$ can be brought into this form

$$(14.5)\qquad \sum_\sigma \sum_\rho \left\{\frac{1}{2}\sum_{i=1}^{N} \mathcal{L}_i^\rho \cdot l_i^\sigma + \sum_{1\le i<j\le N} \mathcal{L}_j^\rho \cdot l_i^\sigma\right\} \quad .$$

A more suitable representation of $\check{C}^{(2)}$ can be obtained via the following identities

$$\begin{aligned} \Big(\sum_{j=1}^{N} \mathcal{L}_j^\rho\Big)\cdot\Big(\sum_{i=1}^{N} l_i^\sigma\Big) &= \sum_{i=1}^{N} \mathcal{L}_i^\rho \cdot l_i^\sigma + \sum_{1\le i<j\le N} \mathcal{L}_j^\rho \cdot l_i^\sigma + \sum_{1\le i<j\le N} \mathcal{L}_i^\rho \cdot l_j^\sigma \\ &= \sum_{i=1}^{N} \mathcal{L}_i^\rho \cdot l_i^\sigma + \sum_{1\le i<j\le N} \mathcal{L}_j^\rho \cdot l_i^\sigma + \sum_{1\le i<j\le N} \mathcal{L}_j^\sigma \cdot l_i^\rho + \\ &\quad + \sum_{1\le i<j\le N} \left(\mathcal{L}_i^\rho \cdot l_j^\sigma - \mathcal{L}_j^\sigma \cdot l_i^\rho\right) \quad . \end{aligned}$$

If all these relations are added up we obtain on the right hand side twice the expression (14.5) plus

$$\sum_\sigma \sum_\rho \sum_{1\le i<j\le N} \left(\mathcal{L}_i^\rho \cdot l_j^\sigma - \mathcal{L}_j^\sigma \cdot l_i^\rho\right) \quad .$$

Note that the general term appearing in the last sum can also be written as Lie-bracket $[l_j^\sigma,\, l_i^\rho]$ (cf. (14.4)).
Resuming the previous notation we are thus arrived at the representation of $\check{C}^{(2)}$ on which all further applications of Theorem 14.1 will be based, namely

$$\overset{\vee}{C}(2) = \frac{1}{2}\sum_{\sigma}\sum_{\rho} \frac{1}{\sigma!\rho!} R_{\sigma,\rho} \quad , \text{ where}$$

$$(14.6) \quad R_{\sigma,\rho} = -\sum_{1\le i<j\le N} [l^{(\sigma)}(u_o, u_{j-1}-u_j), l^{(\rho)}(u_o, u_{i-1}-u_i) z_j^{\sigma} z_i^{\rho} + \left(\sum_{i=1}^{N} l_x^{(\rho)}(u_o, u_{i-1}-u_i) z_i^{\rho}\right)\left(\sum_{i=1}^{N} l^{(\sigma)}(u_o, u_{i-1}-u_i) z_i^{\sigma}\right) .$$

As a first application of the foregoing results we present a lemma and several corollaries which will serve as useful tools in the further course of this paper.

Lemma 14.2 Given $r+1$ different and non-vanishing real numbers $z_1,\dots,z_{r+1}$ and a set of m-dimensional column vectors $c_{\nu,\mu}$, where ν ranges from 1 to r and for every fixed ν the subscript μ ranges from 0 to $\nu-1$. Then one can find sequences

$$V_i = \{v_{i,o}, v_{i,1}, \dots\} , \; i = 1,\dots,r$$

such that for arbitrary u the asymptotic relation

$$C(t,x,\lambda z_1,\dots,\lambda z_{r+1}; u, u+\lambda^r V_1,\dots, u+\lambda^r V_r, u) = x + \lambda^r \left\{\sum_{\nu=1}^{r} \lambda^{\nu}\left(\sum_{\mu=0}^{\nu-1} B_{\mu}(t,x,u) c_{\nu,\mu}\right)\right\} + \mathcal{O}(\lambda^{2r+1})$$

holds (i.e. the remainder term and all its partial derivatives with respect to t and x is of order $\mathcal{O}(\lambda^{2r+1})$, uniformly on compact sets in u).

Proof. From the definition of $C^{(1)}$ as given before it is clear that $C(t,x,\lambda z_1,\dots,\lambda z_{r+1}; u, u+\lambda^r V_1, \dots, +\lambda^r V_r, u) = C^{(1)}(t,x,\lambda z_1,\dots,\lambda z_{r+1}; u, u+\lambda^r V_1, \dots, u+\lambda^r V_r, u) + \mathcal{O}(\lambda^{2r+1})$ hence for the purpose of the proof we may replace the power series C by the series $C^{(1)}$. It follows now from (14.3) and from Theorem 13.1 that the latter one can be written in explicit terms as follows.

$$C^{(1)}(t,x,\lambda z_1,\dots,\lambda z_r;\ \mathcal{U},\ \mathcal{U}+\lambda^r v_1,\ \dots,\ \mathcal{U}+\lambda^r v_r, \mathcal{U}) = x +$$

$$+ \lambda^r \sum_{\nu=1}^{r} \frac{1}{\nu!} \lambda^\nu \sum_{j=0}^{\nu-1} \binom{\nu-1}{j} B_{\nu-1-j}(\mathcal{U}) \sum_{i=1}^{r+1} (v_{i-1,j}-v_{i,j}) z_i^\nu \quad + \mathcal{O}(\lambda^{2r+1})$$

where $v_{o,j} = 0$ for all j. The lemma is therefore proved once we succeed in constructing $v_{i,j}$ with $v_{o,j}=0$ such that the following system of linear equations is satisfied: $v_{r+1,j} = 0$ and

$$\frac{1}{\nu!}\binom{\nu-1}{\nu-1-\mu} \sum_{i=1}^{r+1} (v_{i-1,\nu-1-\mu} - v_{i,\nu-1-\mu}) z_i^\nu = c_{\nu,\mu}$$

for $\nu=1,\dots,r, \mu=0,\dots,\nu-1$. An elementary argument which is described in the appendix (Lemma 1) will indeed establish the unique solvability of this system of equations. Thereby the lemma is proved.

For the remaining portion of this section we consider a fixed triple $t,x,\mathcal{U} = \{u_o,u_1,\dots\}$ and denote by $\mathcal{L} = \mathcal{L}(t,x,\mathcal{U})$ the linear space spanned by the columns of the matrices $B_\nu(t,x,\mathcal{U})$, $\nu=0,1,\dots$

Corollary 1. Assume that $t,x,\mathcal{U}$ satisfy condition (8.19). Then $\mathcal{L}(t,x,\mathcal{U}) \subset P_U(t,x,\mathcal{U})$ (cf. Definition 8.1).

Proof. Given an arbitrary element $p \in \mathcal{L}(t,x,\mathcal{U})$, that is a column vector which can be represented in the form $\sum_{\mu=0}^{r-1} B_\mu(t,x,\mathcal{U})c_\mu$ for some $r>0$ and with some row vectors c_μ.

It follows from the lemma that there exist N real number z_i and sequences $\mathcal{U}_i(\lambda)$ satisfying the relations $z_1<z_2<\dots<z_N<0$ and $C(t,x,\lambda z_1,\dots,\lambda z;\ \mathcal{U},\ \mathcal{U}_1(\lambda),\dots,\ \mathcal{U}_N(\lambda)) = x + \lambda^{2r}p + \mathcal{O}(\lambda^{2r+1})$.
This implies $p \in P_U(t,x,\mathcal{U})$ in view of Definition 8.1.

Corollary 2. Hypothesis (i) r,d,s are positive integers having the properties

$$(14.7)\quad \begin{array}{l} r \leq s \leq 2(r-d), \\ \mathcal{L} \text{ is spanned by the columns of } B_\nu, \nu=0,\ldots,d-1 . \end{array}$$

(ii) $z_i(\lambda)$, $\mathcal{U}_i(\lambda)$, $i=1,\ldots,M$, are scalar functions and $\mathcal{U}$-sequences respectively satisfying conditions (i) and (ii) of Definition 8.1. By $\tilde{C}(\lambda)$ we denote the formal power series of the form (8.18) associated with these quantities, that is

$$(14.8)\quad \tilde{C}(\lambda)=C(t,x,z_1(\lambda),\ldots,z_M(\lambda);\ \mathcal{U},\ \mathcal{U}_1(\lambda),\ldots,\ \mathcal{U}_M(\lambda)) .$$

(iii) The following additional statements are true

$$(14.9)\quad \begin{array}{l} z_M(\lambda) \leq -\gamma\lambda \quad \text{for some positive constant } \gamma, \\ \mathcal{U}_M(\lambda) = \mathcal{U} + \mathcal{O}(\lambda^{s+d-r}) \quad \text{(elementwise)} , \end{array}$$

$$(14.10)\quad \tilde{C}(\lambda) = x + \lambda^r \sum_{\mu=0}^{s-r} \lambda^\mu b_\mu + \lambda^s p + \mathcal{O}(\lambda^{s+1})$$

with $b_\mu \in \mathcal{L}$ for $\mu=0,\ldots,s-r$.

Conclusion: $p \in \mathcal{P}_U(t,x,\mathcal{U})$.

Proof. In order to avoid notational confusion we use the symbols $\bar{x}$ and $\bar{\mathcal{U}}$ to denote the fixed given values of the state vector and the $\mathcal{U}$-sequence. We write x and $\mathcal{U}$ if we wish to indicate that these quantities have to be regarded as variable.
Let us put $r'=r-d$ (this number is positive according to (14.7)) and rewrite (14.10) as follows

$$(14.10')\quad \tilde{C}(\lambda)=\bar{x} + \lambda^{r'} \sum_{\nu=1}^{s-r'} \lambda^\nu b'_\nu + \lambda^s p + \mathcal{O}(\lambda^{s+1})$$

where $b'_\nu \in \mathcal{L}$ for $\nu=1,\ldots,s-r'$, $b'_\nu = 0$ for $\nu=1,\ldots,d-1$. We define b'_ν as zero for $\nu > s-r'$ and determine then column vectors $c_{\nu,\mu}$ such that

$$(14.11)\quad \sum_{\mu=0}^{\nu-1} B_\mu(t,\bar{x},\bar{\mathcal{U}})c_{\nu,\mu} = -b'_\nu , \quad \nu=1,2,\ldots .$$

This is always possible: Take $c_{\nu,\mu} = 0$ for $\nu < d$ and use the fact that for $\nu \geq d$ the linear span of the columns of $B_o, \dots, B_{\nu-1}$ is equal to $\mathcal{L}$.

Next we put $N = M+r'+1$ and choose real numbers z_i and sequences $\mathbf{V}_i$, $i=M+1,\dots,M+r'+1$ such that

(14.12) $\quad -\gamma < z_{M+1} < z_{M+2} < \dots < z_N < 0$

and

(14.13)
$$C(t,x,\lambda z_{M+1},\dots,\lambda z_N;\ \mathbf{U},\ \mathbf{U}+\lambda^{r'}\mathbf{V}_{M+1},\ \dots\dots,\mathbf{U}+\lambda^{r'}\mathbf{V}_{N-1},\mathbf{U}) = x + \lambda^{r'}\Big\{\sum_{\nu=1}^{r'} \lambda^{\nu}\Big(\sum_{\mu=0}^{\nu-1} B_{\mu}(t,x,\mathbf{U})c_{\nu,\mu}\Big\} + \mathcal{O}(\lambda^{2r'+1}) .$$

Here γ is the constant appearing in the first line of (14.8) and $c_{\nu,\mu}$ the column vectors appearing in (14.11). That the relation (14.13) can be established by a proper choice of the $\mathbf{V}_i$ is a consequence of Lemma 14.2. Note that (14.13) holds uniformly in x and $\mathbf{U}$ (in the sense explained in Lemma 14.2). We now specialize x and $\mathbf{U}$ as follows

$$x \to \tilde{C}(\lambda) , \quad \mathbf{U} \to \mathbf{U}_M(\lambda) .$$

Since $s+d-r=s-r'$ and $r'\geq s-r'$ (cf (14.7)) we obtain from (14.9) and (14.10')

$$B_{\mu}(t,\tilde{C}(\lambda),\ \mathbf{U}_M(\lambda)) = B_{\mu}(t,\bar{x},\bar{\mathbf{U}}) + \mathcal{O}(\lambda^{s-r'}) ,$$

and hence, in view of (14.13), (14.11) and (14.10')

$$C(t,\tilde{C}(\lambda),\lambda z_{M+1},\dots,\lambda z_N;\ \mathbf{U}_M(\lambda),\ \mathbf{U}_M(\lambda)+\lambda^{r'}\mathbf{V}_{M+1},\dots\dots,\ \mathbf{U}_M(\lambda)) =$$
$$= \tilde{C}(\lambda)-\lambda^{r'}\sum_{\nu=1}^{s-r'} \lambda^{\nu}b'_{\nu} + \mathcal{O}(\lambda^{s+1}) = \bar{x}+\lambda^{s}p+\mathcal{O}(\lambda^{s+1}) .$$

On the other hand, by virtue of the general composition rule (cf. Theorem 8.1) and in view of (14.8) the left hand side of the above relation can be written in the form

$$C(t,\bar{x},z_1(\lambda),\dots,z_N(\lambda);\ \bar{\mathbf{U}},\ \mathbf{U}_1(\lambda),\dots,\ \mathbf{U}_N(\lambda)) ,$$

and it is not difficult to confirm that all requirements concerning $z_i(\lambda)$ and $\mathcal{U}_i$ listed in Definition 8.1 are met. Note in particular that the inequalities $z_1(\lambda) \leq \ldots \leq z_N(\lambda) \leq 0$ hold because of (14.9) and (14.12). Hence $p \in \mathcal{P}_U(t,\bar{x},\bar{u})$ and the corollary is proved.

15. Transformation of the Control Variable.

In this section we will study the effect which a change of the control variable may have upon the quantities associated with a given system. The type of substitution which we consider is the most general one and given by

$$(15.1) \qquad u \to \hat{u}(t,x,v)$$

where $\hat{u}$ is a m-dimensional vector whose components are C^∞ functions of t,x and a further vector-valued variable v. Except for the differentiability property no restriction is put upon $\hat{u}$, in particular the dimension m' of v is in no way related to the dimension m of u. The transformed system is then

$$(15.2) \qquad \dot{x} = \hat{f}(t,x;v) = f(t,x;\hat{u}(t,x,v))$$

and it is v now which plays the role of the control variable. It will be tacitly assumed that t,x,v are restricted to a subset $\hat{Y}$ of the (t,x,v)-space which is such that $(t,x,v)\in\hat{Y}$ implies $(t,x,\hat{u}(t,x,v))\in Y$. The right hand side of the transformed system (15.2) is then a C^∞ - function of all variables on the set $\hat{Y}$, and hence one can define an operator $\hat{\Gamma} = \Gamma_{\hat{f}}$, matrices $\hat{B}_\nu$ and formal power series $\hat{C}$ with respect to (15.2) in the same way as we have defined Γ, B_ν, C for the system (7.1). We wish to specify the transformation laws for these quantities. To this purpose we introduce a sequence $\mathbf{V}$ of independent variables v_i, that is we put

$$(15.3) \qquad \mathbf{V} = \{v_0, v_1, \ldots\ldots\}$$

in accordance with previous custom. Next we define a sequence

$$\mathbf{U}(t,x,\mathbf{V}) = \{u_o(t,x,\mathbf{V}),u_1(t,x,\mathbf{V})\ldots..\}$$

of functions $u_i(t,x,\mathbf{V})$ recursively as follows

(15.4)
$$u_o(t,x,\mathbf{V}) = \hat{u}(t,x,v_o)$$
$$u_i(t,x,\mathbf{V}) = \tfrac{d}{dt} u_{i-1}(t,x,\mathbf{V}),i=1,2,\ldots,$$

where the differentiation d/dt has to be performed subject to the rules

(15.5)
$$\dot{x} = \hat{f}(t,x;v_o) \quad , \quad \dot{v}_j = v_{j+1} \; , \quad j = 0,1,\ldots$$

It will turn out that - roughly speaking - the substitution (15.1) has to be extended to the substitution $\mathbf{u} \to \mathbf{U}(t,x,\mathbf{V})$ in order to derive the quantities $\hat{B}_v$, $\hat{C}$ from the corresponding quantities B_v, C . Before we can make this more precise we state some simple facts concerning the sequence $\mathbf{U}(t,x,\mathbf{V})$.

<u>Lemma 15.1.</u> (i) $u_i = u_i(t,x,\mathbf{V})$ depends upon $v_o,\ldots,v_i$ only and we have $\delta u_i/\delta v_i = \delta u_o/\delta v_o$. (ii) Let $g(t,x,\mathbf{u})$ be a function of the variables $t,x,\mathbf{u} = \{u_o,u_1,\ldots\}$ and let $\hat{g}$ be defined as
$$\hat{g}(t,x,\mathbf{V}) = g(t,x,\mathbf{U}(t,x,\mathbf{V})) .$$
Claim: The identity

$$\tfrac{d}{dt}\hat{g}(t,x,\mathbf{V}) = \tfrac{d}{dt} g(t,x,\mathbf{u}) \Big|_{\mathbf{u} \to \mathbf{U}(t,x,\mathbf{V})}$$

holds provided differentiation d/dt is performed on the left hand side according to the rules (15.5) and on the right hand side with respect to the system

(15.6) $\dot{x} = f(t,x;u_o)$, $\dot{u}_i = u_{i+1}$, $i=0,1,\ldots$

(iii) Let $v(t)$ be a C^∞ function of t and $x(t)$ a solution of (15.2) for $v=v(t)$. Put $\mathbf{V}(t) = \{v(t),\dot{v}(t),\ldots\}$ and $u(t) = \hat{u}(t,x(t),v(t))$. Then $\{u(t),\dot{u}(t),\ddot{u}(t),\ldots\}= \mathbf{U}(t,x(t),\mathbf{V}(t))$.

<u>Proof.</u> (i) and (iii) can be inferred immediately from the definition (15.4). The rule (ii) will follow for arbitrary g if it has been established in case g=t, g=x, g= u_i and for these

special functions it becomes a trivial consequence of (15.4) .

Theorem 15.1 Let $\Gamma = \Gamma_f$ and $\hat{\Gamma} = \Gamma_{\hat{f}}$ be the operators associated with the systems (7.1) and (15.2) respectively (that is Γ acts on functions $g=g(t,x\ \mathcal{U})$ according to (13.1) and $\hat{\Gamma}$ on functions $h=h(t,x,\mathcal{V})$ according to the formula $\hat{\Gamma}(h)=\partial h/\partial t + \sum(\partial h/\partial v_i)\cdot v_{i+1} + [\hat{f},h]$) and let $\hat{g}(t,x,\mathcal{V}) := g(t,x,\ \mathcal{U}(t,x,\mathcal{V}))$. Then

$$(\hat{\Gamma}(\hat{g}))(t,x,\mathcal{V}) = (\Gamma(g))(t,x,\ \mathcal{U}(t,x,\mathcal{V})) -$$
$$- (\partial f/\partial u)(t,x;u(t,x,v_o))\cdot(\partial\hat{u}/\partial x)(t,x,v_o)\cdot\hat{g}(t,x,\mathcal{V}).$$

Proof. We wish to apply the second definition of the operator Γ given in Sec. 13. To this purpose consider the scalar function

$$y^T\hat{g}(t,x,\mathcal{V}) = y^T g(t,x,\ \mathcal{U}(t,x,\mathcal{V}))$$

and calculate its derivative with respect to the Hamiltonian system

$$(15.7)\qquad \begin{aligned} \dot{x} &= \hat{f}(t,x;v_o) = f(t,x;\hat{u}(t,x,v_o)) \\ \dot{y} &= -\hat{f}_x(t,x;v_o)^T y = -(f_x+f_u\hat{u}_x)^T y\ , \end{aligned}$$

(the argument in f_x and f_u is $t,x,u=\hat{u}(t,x,v_o)$) .
The gives, in view of part (ii) of Lemma 15.1,

$$y^T\cdot \frac{d}{dt}\, g(t,x,\mathcal{U})\Big|_{\mathcal{U}\to\mathcal{U}(t,x,\mathcal{V})} + \dot{y}^T\cdot\hat{g}(t,x,\mathcal{V})\ .$$

Here d/dt denotes differentiation with respect to (15.6). If $\dot{y}$ is replaced by the right hand side of the second equation (15.7) the above expression assumes the form

$$\frac{d}{dt}\,(y^T\cdot g(t,x,\ \mathcal{U}))\Big|_{\mathcal{U}\to\mathcal{U}(t,x,\mathcal{V})} - y^T\cdot(f_u\hat{u}_x\hat{g})\ ,$$

where d/dt now has to be interpreted as differentiation with respect to the Hamiltonian system (13.3). Hence the first term in the last formula can be written as

$$y^T\cdot(\Gamma(g))(t,x,\ \mathcal{U}(t,x,\mathcal{V}))$$

according to (13.2). Comparing what we have obtained so far with the analogue of (13.2) for the operator $\hat{\Gamma}$ (i.e. the relation $d(y^T\hat{g})/dt = y^T \cdot \hat{\Gamma}(\hat{g})$) leads immediately to the desired result.

<u>Corollary 1.</u> Let $B_{..} = B_{..}(t,x,\boldsymbol{u})$ and $\hat{B}_{..} = \hat{B}_{\nu}(t,x,\boldsymbol{v})$ respectively be the matrices of type $n{\times}m$ and $n{\times}m'$ respectively which are defined according to (13.7), where of course in the case of the $\hat{B}_{\nu}$ the role of f and Γ is played by $\hat{f}$ and $\hat{\Gamma}$. Then we have a relation of the form

$$\hat{B}_{\nu}(t,x,\boldsymbol{v}) = B_{\nu}(t,x,\boldsymbol{u}(t,x,\boldsymbol{v}))\cdot(\partial\hat{u}/\partial v)(t,x,v_0) +$$
$$+ \sum_{\rho<\nu} B_{\rho}(t,x,\boldsymbol{u}(t,x,\boldsymbol{v}))\cdot Z_{\nu,\rho}(t,x,\boldsymbol{v})$$

for every $\nu=0,1,\ldots$ The $Z_{\nu,\rho}$ are matrices of type $m{\times}m'$ which consist of C^{∞}- functions of $t,x,\boldsymbol{v}$

<u>Proof.</u> The statement is certainly true in case $\nu=0$ since
$$\hat{B}_0 = \partial\hat{f}/\partial v_0 = (\partial f/\partial u)\cdot(\partial\hat{u}/\partial v) = B_0\cdot(\partial\hat{u}/\partial v) \quad \text{with} \quad v \to v_0 .$$
We proceed by induction and assume that a representation of the said form has already been established for a certain $\hat{B}_{\nu}$. Applying the operator $\hat{\Gamma}$ (columnwise) and using Lemma 13.1, part (iii), one sees that $\hat{\Gamma}(\hat{B}_{\nu}) = \hat{B}_{\nu+1}$ can be written in the form

(15.8) $\quad \hat{\Gamma}(B_{\nu})\cdot\hat{u}_{\nu} + \sum_{\rho<\nu} \hat{\Gamma}(B_{\rho})\cdot\tilde{Z}_{\nu,\rho} + \sum_{\rho\le\nu} B_{\rho}\tilde{\tilde{Z}}_{\nu,\rho}$

where $B_{\mu} = B_{\mu}(t,x,\boldsymbol{u}(t,x,\boldsymbol{v}))$ and the $\tilde{Z}_{\nu,\rho}$ $\tilde{\tilde{Z}}_{\nu,\rho}$ are suitable matrices depending upon $t,x,\boldsymbol{v}$. It follows now from Theorem 15.1 that $\hat{\Gamma}(B_{\mu})$ is equal to $(\Gamma(B_{\mu}))(t,x,\boldsymbol{u}(t,x,\boldsymbol{v}))$ plus a multiple of $B_0(t,x,\boldsymbol{u}(t,x,v_0))$, and thereby one obtains from (15.8) a representation for $\hat{B}_{\nu+1}$ which has the desired form.

<u>Corollary 2.</u> Assume that $\hat{u}(t,x,v)$ does not depend upon $x^1,\ldots,x^{n-1}$ and that the n-th component of g vanishes identically. Then

$$(\hat{\Gamma}(\hat{g}))(t,x,\mathbf{V}) = (\Gamma(g))(t,x,\,\mathcal{U}(t,x,\mathbf{V})).$$

<u>Proof.</u> Follows simply from the theorem and from the observation that - because of the hypothesis - $\hat{u}_x\cdot\hat{g}=0$.

We now come to the central result of this section. It concerns the formal power series which are defined according to Sec. 8 both for the given control system (7.1) and for the transformed system (15.2). The first one is denoted by $C(t,x,z_1,\ldots,z_N;\ \mathbf{U}_o,\ \mathbf{U}_1,\ldots\mathbf{U}_N)$, the second one by $\hat{C}(t,x,z_1,\ldots,z_N;\ \mathbf{V}_o,\ \mathbf{V}_1,\ldots,\ \mathbf{V}_N)$.

<u>Theorem 15.2.</u> The following relation holds identically in $t,x,z_1,\ldots,z_N,\ \mathbf{V}_o\ldots,\ \mathbf{V}_N$:

$$\hat{C}(t,x,z_1,\ldots,z_N;\ \mathbf{V}_o,\ldots,\ \mathbf{V}_N) = \tag{15.9}$$
$$= C(t,x,z_1,\ldots,z_N;\mathbf{U}_o,\ \mathbf{U}_1,\ldots,\ \mathbf{U}_N)\ ,$$

where the $\mathbf{U}_j$ are given in terms of $t,x,z_1,\ldots,z_N,\ \mathbf{V}_o,\ldots,\ \mathbf{V}_N$ as follows ($\mathbf{U}(t,x,\mathbf{V})$ denotes as before the sequence with the components $u_i(t,x,\mathbf{V})$, cf. (15.4)): $\mathbf{U}_o = \mathcal{U}(t,x,\mathbf{V}_o)$,

(15.10) $\mathbf{U}_j = \mathbf{U}(t,\hat{C}(t,x,z_1,\ldots,z_j;\ \mathbf{V}_o,\ldots,\ \mathbf{V}_j),\ \mathbf{V}_j),\ j=1,\ldots,N$.

<u>Remark</u>. In order to obtain the i-th member of the sequence (15.10) one has to take the i-th member of the sequence (15.4) for $\mathbf{V}=\mathbf{V}_j$ and the replace x by the formal power series $\hat{C}(t,x,z_1,\ldots,z_j;\ \mathbf{V}_o,\ldots,\ \mathbf{V}_j)$. That this procedure leads to a well-defined power-series in $z_1,\ldots,z_j$ can be seen by the same type of argument as we have used at earlier occasions (cf. the remark following Theorem 8.1). Furthermore, since the $K_{\underline{\nu}}$ are polynomials in $\mathbf{U}_1,\ldots,\ \mathbf{U}_N$ it is clear that also a well-defined power series in $z_1,\ldots,z_N$ arises out of $K_{\underline{\nu}}$ if each variable $\mathbf{U}_j$ is replaced by the expression on the right hand side of (15.10). Hence rearranging terms will produce again a power

series in $z_1,\dots,z_N$ out of the formal sum $x + \sum(\underline{\nu}!)^{-1}K_{\underline{\nu}}z^{\underline{\nu}}$. According to (8.2) this power series will finally appear on the right hand side of (15.9). It is clear from these considerations that each of its coefficients can be obtained from a finite set of coefficients of $\hat{C}$ by purely algebraic manipulations and hence depends upon finitely many members of each sequence $\mathcal{V}_j$ only.

<u>Proof.</u> We first prove the theorem in case $N=1$. The statement in question runs then as follows

$$(15.11)\qquad \begin{aligned}&\hat{C}(\tilde{t},\tilde{x},z;\ \mathcal{V}_o,\ \mathcal{V}_1) =\\ &= C\Big(\tilde{t},\tilde{x},z;\ \mathcal{U}(\tilde{t},\tilde{x},\ \mathcal{V}_o),\ \mathcal{U}(\tilde{t},\hat{C}(\tilde{t},\tilde{x},z;\ \mathcal{V}_o,\ \mathcal{V}_1),\ \mathcal{V}_1)\Big)\end{aligned}$$

where $z=z_1$ is now scalar. For technical purposes we have written $\tilde{t},\tilde{x}$ instead of t,x; these quantities have to be regarded as fixed throughout the proof of (15.11). It follows from the preceding remark that it suffices again to prove the relation in question in case that at most finitely many elements of each sequence $\mathcal{V}_i$ are different from zero. So without loss of generality we may then assume that there exist two polynomial control function $v_i(t)$ such that

$$(15.12)\qquad \mathcal{V}_i = \{v_i(\tilde{t}),\dot{v}_i(\tilde{t}),\ddot{v}_i(\tilde{t}),\dots\dots\},i=0,1\ .$$

We also assume, according to a previous remark, that

$$(15.13)\qquad (\tilde{t},v_i(\tilde{t})) \in \hat{Y} \text{ and hence } (\tilde{t},\hat{u}(\tilde{t},\tilde{x},v_i(\tilde{t})) \in Y$$

for $i=0,1$.

Let us now consider the differential equations

$$(15.14)\qquad \dot{x} = \hat{f}_i(t,x) = \hat{f}(t,x;v_i(t))\ ,\ i=0,1$$

and let us denote by $\hat{x}_i(t;t_i,a)$ the general solution of (15.14) (for the notion of general solution see Sec. 1). It follows then from Lemma 8.1 and from the formula (4.4) that the left hand side of (15.11) represents the Taylor expansion at $z=0$ of the function $\hat{x}_1\big(\tilde{t};\tilde{t}+z,\hat{x}_o(\tilde{t}+z;\tilde{t},\tilde{x})\big)$. We state this explicitly as our first result, since we have to come back to it later.

(15.15) $\hat{C}(t,\tilde{x},z;\, \mathbf{V}_o, \mathbf{V}_1)$ is the Taylor-expansion at $z=0$ of the function $\hat{x}_1\big(\tilde{t};\tilde{t}+z,\hat{x}_o(\tilde{t}+z;\tilde{t},\tilde{x})\big)$.

Our next aim is to construct a function whose Taylor-expansion at $z=0$ is given by the series on the right hand side of (15.11). To this purpose let us consider the differential equations

(15.16) $\dot{x} = f_i^*(t,x;\tau,b) = f\big(t,x;\hat{u}(t,\hat{x}_i(t;\tau,b),v_i(t))\big)$

$i=0,1$. The scalar τ and the n-dimensional vector b have to be regarded as parameters. The f_i^* are of class C^∞ on some neighborhood of the point $t = \tau = \tilde{t}$, $x = b = \tilde{x}$, because of (15.13) . Hence the general solution of (15.16), which will be denoted by

$$x_i^*(t;t_o,a,\tau,b) ,$$

is also of class C^∞ in a neighborhood of the point whose coordinates satisfy $t = t_o = \tau = \tilde{t}$, $a = b = \tilde{x}$. Therefore the function

(15.17) $x_1^*\big(\tilde{t};\tilde{t}+z,x_o^*(\tilde{t}+z;\tilde{t},\tilde{x},\tau_o,b_o),\tau_1,b_1\big)$

is well defined and of class C^∞ if the variables z,τ_o,τ_1, b_o,b_1 are restricted to a neighborhood of the point $(0,\tilde{t},\tilde{t},\tilde{x},\tilde{x})$. Using the relation (4.4) and applying Lemma 8.1 once more we find that the Taylor-expansion of (15.17) at $z=0$ is given by $C(\tilde{t},\tilde{x},z;\, \mathbf{U}_o^*, \mathbf{U}_1^*)$ where

$$\mathbf{U}_i^* = \{\hat{\hat{u}}_i(\tilde{t}),((\partial/\partial t)\hat{\hat{u}}_i)(\tilde{t}),((\partial^2/\partial t^2)\hat{\hat{u}}_i)(\tilde{t}),\dots\}$$

and $\hat{\hat{u}}_i(t) = \hat{u}(t,\hat{x}_i(t;\tau_i,b_i),v_i(t))$.

Because of (15.12) and because of the fact that $\hat{x}_i$, regarded as a function of t, is a solution of the eq. (15.14), the sequence $\mathbf{U}_i^*$ coincides with $\mathbf{U}(\tilde{t},\hat{x}_i(\tilde{t};\tau_i,b_i),\, \mathbf{V}_i)$. This can be seen from part (iii) of Lemma 15.1. Hence we are arrived at our second result: The power series

$$C\big(\tilde{t},\tilde{x},z;\; \mathbf{U}(\tilde{t},\hat{x}_o(\tilde{t};\tau_o,b_o),\mathbf{V}_o),\; \mathbf{U}(\tilde{t},\hat{x}_1(\tilde{t};\tau_1,b_1),\, \mathbf{V}_1)\big)$$

is the Taylor-expansion of the function (15.17) at $z=0$. In view of this result it is now rather obvious how one can find a function whose expansion at $z=0$ is given by the power series on the right hand side of the relation (15.11). The function (15.17) will exhibit this property provided the parameters τ_i, b_i are replaced by C^∞ functions of z which are such that

$$\hat{x}_o(\tilde{t};\tau_o,b_o) = \tilde{x}\ ,\quad \hat{x}_1(\tilde{t};\tau_1,b_1) \sim \hat{C}(\tilde{t},\tilde{x},z;\mathbf{V}_o\mathbf{V}_1).$$

To fulfill the last condition is easy in view of (15.15). Take

$$\text{(15.18)}\qquad \begin{aligned} \tau_o &= \tilde{t}\ , & b_o &= \tilde{x} \\ \tau_1 &= \tilde{t}+z\ , & b_1 &= \hat{x}_o(\tilde{t}+z;\tilde{t},\tilde{x}). \end{aligned}$$

In other words: If the function which arises from (15.17) by specializing τ_i and b_i according to (15.18) is expanded at $z=0$ one obtains just the power series which appears on the right hand side of (15.11). Combining this statement with (15.15) leads to the guess that the background of (15.11) lies in the following fact: If the substitution (15.18) is carried out in (15.17) the result will be the function $\hat{x}_1\big(\tilde{t};\tilde{t}+z,\hat{x}_o(\tilde{t}+z;\tilde{t},\tilde{x})\big)$. This indeed is correct and can be derived almost immediately from the relation

$$x_i^*(t;t_o,a,t_o,a) = \hat{x}_i(t;t_o,a)$$

which in turn is a simple consequence of the fact that $\hat{x}_i(t;\tau,b)$- regarded as a function of t - is a solution of the differential eq. (15.16). Thereby the statement of the theorem is proved in case $N=1$. The remaining portion of the proof is handled by induction. We establish the assertion for an arbitrary N under the assumption that the theorem has already been proved for all N' such that $1 \leq N' < N$. To begin with we apply the general composition rule with $M=2$ to the left hand side of (15.9) and obtain

$$\hat{C}(t,\hat{C}(t,x,z_1;\mathbf{V}_o,\mathbf{V}_1),\ z_2,\ldots,z_N;\ \mathbf{V}_1,\ldots,\ \mathbf{V}_N)\ .$$

Using hypothesis of induction this can be changed to

(15.19) $\quad C(t,x', z_2,\dots,z_N, \mathcal{U}_1',\dots, \mathcal{U}_N')$, where

$$x' = \hat{C}(t,x,z_1; \mathcal{V}_o, \mathcal{V}_1), \quad \mathcal{U}_1' = \mathcal{U}(t,x',\mathcal{V}_1) \text{ and}$$

$$\mathcal{U}_i' = \mathcal{U}(t,\hat{C}(t,x',z_2,\dots,z_i; \mathcal{V}_1,\dots, \mathcal{V}_i), \mathcal{V}_i),\ i=2,\dots,N.$$

If we express x' in terms of x and use the general composition rule again we obtain

(15.20) $\quad \mathcal{U}_i' = \mathcal{U}(t,\hat{C}(t,x,z_1,\dots,z_i; \mathcal{V}_o,\dots, \mathcal{V}_i), \mathcal{V}_i)$

for $i=1,\dots,N$. Finally we have, in view of (15.11),

$$x' = C(t,x,z_1; \mathcal{U}(t,x, \mathcal{V}_o), \mathcal{U}_1')$$

and therefore, using the general composition rule once more, (15.19) can be rewritten as

$$C(t,x,z_1,\dots,z_N; \mathcal{U}(t,x, \mathcal{V}_o), \mathcal{U}_1',\dots, \mathcal{U}_N') .$$

This however is, in view of (15.20), nothing else than the series on the right hand side of (15.9) and thereby this relation has been established in full generality.

Corollary 1. Let Q be a constant matrix and $\hat{u}(t,x,v) = Qv$. Then $\mathcal{U}(t,x,\mathcal{V}) = Q\mathcal{V} = \{Qv_o, Qv_1,\dots\}$ and

$$\hat{C}(t,x,z_1,\dots,z_N; \mathcal{V}_o,\dots, \mathcal{V}_N) =$$
$$= C(t,x,z_1,\dots,z_N; Q\mathcal{V}_o,\dots,Q\mathcal{V}_N) .$$

Proof. Follows immediately from (15.4) and from Theorem 15.2.

Corollary 2. Let the following condition be satisfied

(i) f does not depend upon x^n, $\hat{u}(t,x,v)$ does not depend

(15.21) upon $x^1,\dots,x^{n-1}$ (note that $x=(x^1,\dots,x^n)^T$) ,

(ii) f^n(=n-th component of f) is constant.

Then the statement of Theorem 15.2 can be simplified as follows

$$\hat{C}(t,x,z_1,\dots,z_N; \mathcal{V}_o,\dots,\mathcal{V}_N)=$$
$$= C(t,x,z_1,\dots,z_N; \mathcal{U}(t,x, \mathcal{V}_o), \mathcal{U}(t,x, \mathcal{V}_1),\dots, \mathcal{U}(t,x,\mathcal{V}_N)) .$$

Proof. The n-th component $\hat{f}^n$ of the transformed equation is constant again, hence the n-th component of $\hat{C}(t,x,\dots)$ reduces to x^n (cf. Lemma 8.2). Furthermore $\mathcal{U}(t,x,\mathcal{V})$ does not depend upon $x^1,\dots,x^{n-1}$. This is true, because of the hypothesis of the corollary, for the first member of this sequence. It follows for the subsequent members simply from the observation that - because of $\hat{f}^n$ = const - differentiation with respect to the eq. $\dot{x}=\hat{f}(t,x;v)$ cannot create dependence from $x^1,\dots,x^{n-1}$, if such a dependence was not there before. Hence $\mathcal{U}(t,x,\mathcal{V})$ remains unchanged if x is replaced by some $\hat{C}(t,x,\dots)$.

We conclude this section by presenting a further application of Theorem 15.2 which could be called an invariance principle. It states, roughly speaking, that sets $\mathcal{P}_U$, which were defined in Sec. 8 and which enter into the formulation of the higher order necessary conditions, remain invariant under the substitution $\mathcal{U}\to\mathcal{U}(t,x,\mathcal{V})$.

Corollary 3. (Invariance Principle). Given fixed t,x and subsets U,V of the u- and v-space respectively such that $v \in \text{int } V$ implies $\hat{u}(t,x,v) \in \text{int } U$. Then the relation

$$\hat{\mathcal{P}}_V(t,x,\mathcal{V}) \subseteq \mathcal{P}_U(t,x,\mathcal{U}(t,x,\mathcal{V})) \tag{15.22}$$

holds provided $\mathcal{V} = \{v_o,v_1,\dots\}$ satisfies the condition

$$(t,x,v_o) \in \hat{Y}, \quad v_o \in \text{int } V .$$

Here $\mathcal{P}$, $\hat{\mathcal{P}}$ respectively are the sets associated - in the sense of Definition 8.1 - with the systems (7.1) and (15.2) respectively.

Proof. Given $p \in \hat{\mathcal{P}}_V$. We then determine integers s,N, scalars $z_i(\lambda)$ and sequences $\mathcal{V}_i(\lambda) = \{v_{i,o}(\lambda),\dots\}$ such that all conditions listed in Definition 8.1 are satisfied and the asymptotic relation

$$\begin{aligned}\hat{C}(t,x,z_1(\lambda),\dots,z_N(\lambda);\ \mathcal{V},\ \mathcal{V}_1(\lambda),\dots,\ \mathcal{V}_N(\lambda)) &= \\ = x+\lambda^s p+\ \mathcal{O}(\lambda^s)&\end{aligned} \tag{15.23}$$

holds. We apply the Theorem 15.2 and change the left hand side of (15.23) to a power series of the C-type. This leads to the relation

$$C(t,x,z_1(\lambda),\dots,z_N(\lambda);\mathcal{U},\mathcal{U}_1(\lambda),\dots,\mathcal{U}_N(\lambda))=x+\lambda^s p+\mathcal{O}(\lambda^s) \tag{15.24}$$

where $\mathcal{U} = \mathcal{U}(t,x,\mathcal{V})$ and

$$\mathcal{U}_i(\lambda)=\mathcal{U}\left(t,\hat{C}(t,x,z_1(\lambda),\dots,z_i(\lambda);\mathcal{V},\mathcal{V}_1(\lambda),\dots,\mathcal{V}_i(\lambda)),\mathcal{V}_i(\lambda)\right),$$

for $i=1,\dots,N$. $\mathcal{U}_i(\lambda)$ is a formal power series in λ, the λ-free term being $\mathcal{U}_i(0)=\mathcal{U}(t,x,\mathcal{V}_i(0))=\{\hat{u}(t,x,v_{i,o}(0)),\dots\}$. Since it is required, according to Definition 8.1, that $v_{i,o}(0)$ satisfy the conditions

$$(t,x,v_{i,o}(0))\in\hat{Y} \quad\text{and}\quad v_{i,o}(0)\in \operatorname{int} V ,$$

it is clear, by virtue of the hypothesis of the **corollary** that

$$\left(t,x,\hat{u}(t,x,v_{i,o}(0))\right)\in Y \quad\text{and}\quad \hat{u}(t,x,v_{i,o}(0))\in \operatorname{int} U .$$

It is now easy, in view of the first remark following Definition 8.1, to infer from (15.24) the conclusion of the theorem, namely $p\in \mathcal{P}_U$.

16. Some identities involving Lie-brackets and the operator Γ.

The purpose of this section is to establish certain formal relations which will be needed in the sequel. We begin with two simple identities which concern repeated Lie-brackets formed out of arbitrary sufficiently often differentiable column vectors a,b. These relations are stated as our first lemma and comprise all background material about Lie-algebras required for the remaining portion of this paper. We adopt the notation introduced in Sec. 13 (cf. in particular (13.6)).

Lemma 16.1. The following identity holds for $\mu\ge 0$, $\nu\ge 0$

$$(-1)^{\mu}[\mathrm{ad}^{\mu}(a)b,\mathrm{ad}^{\nu}(a)b] = \sum_{\tau=0}^{\mu}\binom{\mu}{\tau}(-1)^{\tau}(\mathrm{ad}^{\tau}(a)[b,\mathrm{ad}^{\mu+\nu-\tau}(a)b]) .$$

Proof. The statement is obtained from the binomial formula

$$(16.1)\quad (-1)^{\mu}[\mathrm{ad}^{\mu}(a)b,c] = \sum_{\tau=0}^{\mu}\binom{\mu}{\tau}(-1)^{\tau}(\mathrm{ad}^{\tau}(a)[b,\mathrm{ad}^{\mu-\tau}(a)c])$$

if one specializes c to $ad^{\nu}(a)b$. (16.1) is trivial for $\mu=0$ and reduces to the Jacobi identity

(16.1') $-[[a,b],c] = [b,[a,c]] - [a,[b,c]]$

in case $\mu=1$. The general case is proved by induction using (16.1') with $ad^{\mu}(a)b$ instead of b.

Thereby the lemma is proved.

For the remaining portion of this section we make the following assumption with respect to the underlying control system (7.1)

(16.2) $f = f(x;u)$ does not depend explicitly upon t,
u is scalar .

Note that under this condition also the quantities $B_{\nu}, l^{(\nu)}, L^{(\nu)}$ do not depend upon t. The B_{ν} are column vectors, hence the Lie-brackets $[B_{\mu}, B_{\nu}]$ are well defined.

Lemma 16.2. The identity

$$[B_{\mu}, B_{\nu}] = (-1)^{\mu} \sum_{\tau=0}^{\mu+1} \binom{\mu+1}{\tau} (-1)^{\tau} \Gamma^{\tau} (\partial B_{\nu+\mu+1-\tau}/\partial u_o)$$

holds for all $\nu \geq 0$, $\mu \geq 0$ on the set

(16.3) $\{x, U = \{u_o, u_1, \ldots\} : u_i = 0 \text{ for } i > 0\}$.

Here Γ^{τ} denotes the τ-th iterate of the operator Γ .

Proof. Lemma 16.1 (with $a = f(x;u_o), b = B_o$) and Lemma 13.1 (part (i) and the corollary) yield the following relations:

$B_{\mu} = ad^{\mu}(a)b$, $\mu = 0,1,\ldots,$ and

(16.4) $(-1)^{\mu}[B_{\mu}, B_{\nu}] = \sum_{\tau=0}^{\mu} \binom{\mu}{\tau} (-1)^{\tau} \Gamma^{\tau} ([B_o, B_{\mu+\nu-\tau}])$

which hold true on the set (16.3). We now have, in view of (13.7) and the second of the relations (13.4),

$$[B_o, B_{\mu+\nu-\tau}] = \partial B_{\mu+\nu+1-\tau}/\partial u_o - \Gamma(\partial B_{\mu+\nu-\tau}/\partial u_o) ,$$

and the right hand side of (16.4) can hence be written in this form

$$\sum_{\tau=0}^{\mu} \binom{\mu}{\tau}(-1)^{\tau}\Big(\Gamma^{\tau}(\partial B_{\mu+\nu+1-\tau}/\partial u_0) - \Gamma^{(\tau+1)}(\partial B_{\mu+\nu-\tau}/\partial u_0)\Big) =$$

$$= \sum_{\tau=0}^{\mu} \binom{\mu}{\tau}\Big((-1)^{\tau}\Gamma^{\tau}(\partial B_{\mu+\nu+1-\tau}/\partial u_0)+(-1)^{\tau+1}\Gamma^{(\tau+1)}(\partial B_{\mu+\nu-\tau}/\partial u_0)\Big)$$

$$= \sum_{\tau=0}^{\mu+1} \binom{\mu+1}{\tau}(-1)^{\tau}\Gamma^{\tau}\Big(\partial B_{\nu+\mu+1-\tau}/\partial u_0\Big) .$$

Thereby the lemma is proved.

<u>Corollary.</u> Let $\mathbf{o}$, $\mathbf{e}$ respectively denote the sequences $\{0,0,\ldots\}$ and $\{1,0,0,\ldots\}$ respectively. Then, for $\nu \geq 1$,

$$L^{(\nu)}(\mathbf{o},\mathbf{e})\cdot\mathbf{e} + \sum_{\substack{\rho+\sigma=\nu \\ \rho,\sigma>0}} \frac{(\nu-1)!}{\rho!(\sigma-1)!}[1^{(\sigma)}(\mathbf{o},\mathbf{e}),1^{(\rho)}\mathbf{o},\mathbf{e})] =$$

$$= \Big(\Gamma^{(\nu-1)}(\partial B_0/\partial u_0)\Big)(\mathbf{o}) .$$

Here again the argument x has been omitted.

<u>Proof.</u> It follows from Theorem 13.1 and from Definition 12.1 that

$$(16.5)\quad 1^{(\sigma)}(\mathbf{o},\mathbf{e}) = B_{\sigma-1}(\mathbf{o}),\ L^{(\nu)}(\mathbf{o},\mathbf{e})\cdot\mathbf{e}=(\partial B_{\nu-1}/\partial u_0)(\mathbf{o}) .$$

From now on we omit also the arguments $\mathbf{o},\mathbf{e}$. Combining (16.5) with Lemma 16.2 we obtain, for $\nu \geq 2$ this relation

$$\sum_{\substack{\rho+\sigma=\nu \\ \rho,\sigma>0}} \frac{1}{\rho!(\sigma-1)!}[1^{(\rho)},1^{(\sigma)}] = \sum_{\rho+\sigma=\nu-2} \frac{1}{(\rho+1)!\sigma!}[1^{(\rho+1)},1^{(\sigma+1)}] =$$

$$= \sum_{\rho+\sigma=\nu-2} \frac{1}{(\rho+1)!\sigma!}[B_{\rho},B_{\sigma}]$$

$$= \sum_{\rho+\sigma=\nu-2} \frac{(-1)^{\rho}}{(\rho+1)!\sigma!} \sum_{\tau=0}^{\rho+1} \binom{\rho+1}{\tau}(-1)^{\tau}\Gamma^{\tau}(\partial B_{\nu-1-\tau}/\partial u_0)$$

$$= \sum_{\tau=0}^{\nu-1} \alpha_{\nu,\tau} \frac{1}{\tau!} \Gamma^{\tau}\left(\partial B_{\nu-1-\tau}/\partial u_o\right)$$

Here $\alpha_{\nu,\tau}$ denotes the rational number

$$\sum \frac{(-1)^{\rho-\tau}}{\sigma!(\rho+1-\tau)!} = \frac{1}{(\nu-1-\tau)!} \sum \binom{\nu-1-\tau}{\rho+1-\tau} (-1)^{\rho-\tau}$$

where the summation has to be extended over all ρ, σ such that

$$\rho+\sigma = \nu-2 \geq 0 \ , \ \rho+1 \geq \tau \ , \ \rho \geq 0 \ .$$

Putting $\rho = \tau-1+\omega$, $\sigma = \nu-1-\tau-\omega$ the above sum can be brought into this form

$$- \frac{1}{(\nu-1-\tau)!} \sum_{\omega=\delta}^{\nu-1-\tau} (-1)^{\omega} \binom{\nu-1-\tau}{\omega}$$

where $\delta=0$ for $\tau > 0$ and $\delta=1$ for $\tau = 0$. Hence we have

$$\alpha_{\nu,0} = \frac{1}{(\nu-1)!} \ ; \ \alpha_{\nu,\tau} = 0 \ \text{ for } \ 0 < \tau < \nu-1 \ , \ \alpha_{\nu,\nu-1} = -1$$

and

$$\sum_{\substack{\rho+\sigma=\nu \\ \rho,\sigma>0}} \frac{1}{\rho!(\sigma-1)!} [1^{(\rho)}, 1^{(\sigma)}] = \frac{1}{(\nu-1)!}\left(\partial B_{\nu-1}/\partial u_o - \Gamma^{(\nu-1)}(\partial B_o/\partial u_o)\right) .$$

This is nothing else than the statement in question as can be seen from (16.5). Thereby the corollary has been proved in case $\nu>1$, the case $\nu=1$ is an immediate consequence of (16.5) .

17. A general formula for the Lie-brackets $[B_{\mu}^{\rho} , B_{\nu}^{\sigma}]$.

The identity expressed in Lemma 16.2 admits certain generalizations which are fundamental for the remaining portion of this work and which are stated as Theorem 17.1 and 17.2. The second one resembles very much our previous result, however there are important differences. The argument in B.. ,∂B.../∂u_o is arbitrary and not restricted to the set (16.3). Secondly u need not to be scalar, hence the quantities occuring in the formula are not the matrices B_{μ} but their columns which will be numerated by means of superscripts and hence denoted by B_{μ}^{σ} , $\sigma=1,\dots,m$. **note that** $B_o^{\sigma}=\partial f/\partial u_o^{\sigma}$

The proof of the theorem requires a certain amount of preparatory considerations into which we enter now. We resume the notation introduced in Sec. 14. The integer N will now always be equal to 2, hence the multiindex $\underline{\nu}$ is actually a pair (σ,ρ) of integers. We write then, as before, $K_{(\sigma,\rho)}, K^{(\lambda)}_{(\sigma,\rho)}$, instead of $K_{\underline{\nu}}, K^{(\lambda)}_{\underline{\nu}}$.

Lemma 17.1. The relation

$$\sum_{\sigma+\rho=\nu} \frac{1}{\sigma!\rho!} K_{(\sigma,\rho)}(\mathbf{u}_o, \mathbf{u}_1, \mathbf{u}_o) z_1^{\sigma} z_2^{\rho} =$$

$$= \sum_{\sigma+\rho=\nu} \frac{1}{\sigma!\rho!} K^{(2)}_{(\sigma,\rho)}(\mathbf{u}_o, \mathbf{u}_1, \mathbf{u}_o) z_1^{\sigma} z_2^{\rho} + \mathcal{O}(|z_1-z_2| \|\mathbf{u}_1-\mathbf{u}_o\|^3)$$

holds for every $\nu > 0$ in the following sense. The remainder term and all partial derivatives with respect to the (omitted) variables t,x satisfies an estimate of the form $\leq \text{const}|z_1-z_2| \|\mathbf{u}_1-\mathbf{u}_o\|^3$ on compact sets.

Proof. Using part (ii) of Lemma 7.4 and the fact that $K^{(2)}_{\underline{\nu}}$ is the second order Taylor polynomial of $K_{\underline{\nu}}$ we obtain the identity for $\nu > 0$

$$\sum_{\sigma+\rho=\nu} \frac{1}{\sigma!\rho!}(K_{(\sigma,\rho)}(\mathbf{u}_o,\mathbf{u}_1,\mathbf{u}_o)-K^{(2)}_{(\sigma,\rho)}(\mathbf{u}_o,\mathbf{u}_1,\mathbf{u}_o)) = 0 .$$

If this relation is multiplied with $z_1^{\nu} = z_1^{\sigma} z_2^{\rho} + \mathcal{O}(|z_1-z_2|)$ we obtain the desired result, in view of the fact that

$$K_{(\sigma,\rho)}-K^{(2)}_{(\sigma,\rho)} = \mathcal{O}(\|\mathbf{u}_1-\mathbf{u}_o\|^3) .$$

In what follows $t,x,\mathbf{u} = \{u_o,u_1,\dots\}$, $\mathbf{v} = \{v_o,v_1,\dots\}$ and the scalar λ have to be regarded as independent variables, $\mathbf{u}_i, z_i$, as functions of $\lambda,\mathbf{u},\mathbf{v}$. The arguments t,x again are omitted here and there.

Lemma 17.2 Let r be a positive integer and let $\mathcal{U}_i, z_i$ be chosen as follows

$$\begin{aligned} &\mathcal{U}_0 = \mathcal{U}_2 = \mathcal{U}, \quad \mathcal{U}_1 = \mathcal{U} + \lambda^r \mathcal{V} \;; \\ &z_1 = \lambda, \; z_2 = \lambda + \lambda^r . \end{aligned} \tag{17.1}$$

Claim: A relation of the form

$$C(t,x,z_1,z_2;\; \mathcal{U}_0, \mathcal{U}_1, \mathcal{U}_2) =$$

$$= x + \lambda^{2r} \sum_{\rho=1}^{2r} \lambda^{\rho-1} c_\rho + \frac{1}{2} \lambda^{3r} \sum_{\nu=1}^{r} \lambda^{\nu-1} g_\nu + \mathcal{O}(\lambda^{4r})$$

holds uniformly in $t, x, \mathcal{U}, \mathcal{V}$ (on compact sets). The coefficients c_ρ are linear combinations of $l^{(\mu)}(t,x;\mathcal{U},\mathcal{V})$ with integers as coefficients. g_ν is a quadratic form in $\mathcal{V}$ and given as

$$g_\nu = g_\nu(\mathcal{U},\mathcal{V}) := \frac{1}{(\nu-1)!} L^{(\nu)}(\mathcal{U},\mathcal{V})\cdot\mathcal{V} \tag{17.2}$$

$$+ \sum_{\substack{\rho+\sigma=\nu \\ \rho,\sigma>0}} \frac{1}{\rho!(\sigma-1)!} [l^{(\sigma)}(\mathcal{U},\mathcal{V}), l^{(\rho)}(\mathcal{U},\mathcal{V})] .$$

Proof. It follows from Lemma 17.1 and from (17.1) that $C(z_1,z_2;\mathcal{U}_0,\mathcal{U}_1,\mathcal{U}_0) = C^{(2)}(z_1,z_2;\mathcal{U}_0,\mathcal{U}_1,\mathcal{U}_0) + \mathcal{O}(\lambda^{4r})$, hence for the purpose of the proof we may replace C by $C^{(2)}$ and use the representation for the latter one which was derived in Sec. 14. First we note that, in view of (17.1)

$$\begin{aligned} \sum_{i=1}^{2} l^{(\nu)}(\mathcal{U}_0, \mathcal{U}_{i-1} - \mathcal{U}_i) z_i^\nu &= \lambda^r \{-l^{(\nu)}(\mathcal{U},\mathcal{V})\lambda^\nu + l^{(\nu)}(\mathcal{U},\mathcal{V})(\lambda+\lambda^r)^\nu\} \\ &= \lambda^{2r} \pi_\nu(\lambda) l^{(\nu)}(\mathcal{U},\mathcal{V}) = \mathcal{O}(\lambda^{2r}) , \end{aligned} \tag{17.3}$$

where $\pi_\nu(\lambda)$ is a polynomial in λ with integer coefficients. It follows therefore from (14.3) that

$$C^{(1)}(z_1,z_2;\mathcal{U}_0,\mathcal{U}_1,\mathcal{U}_0) = x + \lambda^{2r} \sum_{\rho=1}^{\infty} \lambda^{\rho-1} c_\rho \tag{17.4}$$

where the c_ρ are as specified above. Furthermore we have, in view of (14.6) and (17.3),

$$R_{\sigma,\rho} = - [l^{(\sigma)}(\mathbf{u}_o, \mathbf{u}_1 - \mathbf{u}_2), l^{(\rho)}(\mathbf{u}_o, \mathbf{u}_o - \mathbf{u}_1)] z_2^\sigma z_1^\rho + \mathcal{O}(\lambda^{4r}) =$$
$$= \lambda^{2r}[l^{(\sigma)}(\mathbf{u}, \mathbf{v}), l^{(\rho)}(\mathbf{u}, \mathbf{v})]\lambda^\rho(\lambda+\lambda^r)^\sigma + \mathcal{O}(\lambda^{4r}) =$$
$$= \lambda^{2r}(\lambda^{\rho+\sigma} + \sigma\lambda^{\rho+\sigma-1}\lambda^r)[l^{(\sigma)}(\mathbf{u},\mathbf{v}), l^{(\rho)}(\mathbf{u},\mathbf{v})] + \mathcal{O}(\lambda^{4r}) .$$

It follows now from the skew symmetry of the Lie-bracket that

$$\sum_{\sigma,\rho>0} \frac{\lambda^{\rho+\sigma}}{\sigma!\rho!}[l^{(\sigma)}, l^{(\rho)}] = 0$$

and we infer therefore from the above representation of $R_{\sigma,\rho}$ that

$$(17.5) \quad \check{C}^{(2)} = \frac{1}{2}\lambda^{3r}\sum_{\nu=1}^{r} \lambda^{\nu-1} \sum_{\substack{\sigma+\rho=\nu \\ \sigma,\rho>0}} \frac{1}{\rho!(\sigma-1)!}[l^{(\sigma)}, l^{(\rho)}]$$

plus terms of order $\mathcal{O}(\lambda^{4r})$. Finally we turn to the study of

$$\sum_{i=1}^{2} L^{(\nu)}(\mathbf{u}_o, \mathbf{u}_{i-1} - \mathbf{u}_i)\cdot(\mathbf{u}_{i-1} + \mathbf{u}_i - 2\mathbf{u}_o) z_i^\nu \quad .$$

Using for $\mathbf{u}_i, z_i$ the special values given by (17.1) this expression can be written as

$$\lambda^{2r}\{\lambda^\nu L^{(\nu)}(\mathbf{u}, -\mathbf{v})\cdot\mathbf{v} + (\lambda+\lambda^r)^\nu L^{(\nu)}(\mathbf{u},\mathbf{v})\cdot\mathbf{v}\}$$
$$= \lambda^{3r}\nu\lambda^{\nu-1}L^{(\nu)}(\mathbf{u}, \mathbf{v})\cdot\mathbf{v} + \mathcal{O}(\lambda^{4r}) \quad .$$

Combining the last result with (17.3), (17.4) one obtains from (14.3) the statement of the lemma.

Occasionally we assume that the function f on the right hand side of the underlying system (7.1) does not depend explicitly upon t . This implies that B_ν, $l^{(\nu)}$, $L^{(\nu)}$ and hence also the column vectors $g_\nu = g_\nu(x, \mathbf{u}, \mathbf{v})$, which we have introduced by (17.2), do not depend explicitly upon t.

<u>Theorem 17.1.</u> The following relation holds identically in x, y, $\mathbf{u} = \{u_o, u_1, \dots\}$, $\mathbf{v} = \{v_o, v_1, \dots\}$ ($u_o^{(\sigma)}, v_o^{(\sigma)}$ are the components of $u_o \cdot v_o$).

$$(17.6)\quad \sum_{\sigma,\rho=1}^{m} \frac{d^{\nu-1}}{dt^{\nu-1}}\Big((y^T\cdot\frac{\partial B_o^\rho}{\partial u_o^{(\sigma)}}(x,\mathbf{U})v_o^{(\sigma)}v_o^{(\rho)}\Big) = (\nu-1)!\,y^T\cdot g_\nu(x,\mathbf{U},\mathbf{V})\ ,$$

where d/dt means differentiation with respect to the system

$$\dot{x} = f(t,x;u_o),\ \dot{y} = -f_x(t,x;u_o)^T y,\ \dot{u}_i = u_{i+1},\ \dot{v}_i = v_{i+1}\quad .$$

Proof. It is clear, in view of (13.2), that the left hand side of (17.6) is the scalar product of y and a vector which can be represented in the form

$$(17.7)\quad \sum_{\lambda=0}^{\nu-1}\ \sum_{\sigma,\rho=1}^{m} \binom{\nu-1}{\lambda}\pi_{\sigma,\rho}^{(\lambda)}\Big(\Gamma^\lambda(\partial B_o^\rho/\partial u_o^{(\sigma)})\Big)(x,\mathbf{U})\ .$$

Here $\pi_{\sigma,\rho}^{(\lambda)}$ is a scalar quadratic form in $\mathbf{V}$ which in explicit terms can be written as

$$(17.8)\quad \pi_{\sigma,\rho}^{(\lambda)}(\mathbf{V}) = \frac{d^{\nu-1-\lambda}}{dt^{\nu-1-\lambda}}(v_o^{(\sigma)}v_o^{(\rho)})$$

if differentiation d/dt is carried out according to the rule $\dot{v}_i = v_{i+1}$. The result which will come out of our considerations at the end of the proof is an identity of the form

$$(17.9)\qquad \sum_\lambda\sum_{\sigma,\rho} \binom{\nu-1}{\lambda}\pi_{\sigma,\rho}^{(\lambda)}\Big(\Gamma^\lambda(\partial B_o^\rho/\partial u_o^{(\sigma)})\Big) + \tilde{g}_\nu = (\nu-1)!\,g_\nu$$

where $\tilde{g}_\nu$ is linear in $\mathbf{V}$. Comparing quadratic terms on both sides of (17.9) we conclude that $\tilde{g}_\nu$ actually is zero and therefore the lemma holds true in the form as stated above.

Since - for fixed ν - all quantities involved depend upon finitely many components of $\mathbf{U}$, $\mathbf{V}$ only we may assume without loss of generality that all but finitely many components of $\mathbf{U}$, $\mathbf{V}$ are zero. We also may assume that the right hand side $f = (f^1,\dots,f^n)^T$ of the underlying system satisfies the additional hypothesis

(17.10) f does not depend upon x^n. $f^n \equiv 1$.

Indeed, if the lemma has been proved under the hypotheses (17.10) it has in fact been proved for the augmented system (13.11). The validity of our statement for the original system is then an immediate consequence of Lemma 13.3 and the subsequent remarks. Whenever $\mathbf{U}, \mathbf{V}, x$ have to be regarded as fixed throughout the proof they are denoted by the symbols $\bar{\mathbf{U}}, \bar{\mathbf{V}}, \bar{x}$ in order to avoid notational confusion *), in particular $\bar{x}^{(n)}$ (=n-th component of $\bar{x}$) is a fixed number)

Let $\tilde{u}(\xi),\tilde{v}(\xi) = (\tilde{v}^{(1)}(\xi),\ldots,\tilde{v}^{(m)}(\xi))^T$ be two m-dimensional vectors whose components are polynomials in a scalar variable ξ and which are such that

$$(17.11)\qquad \begin{aligned} \bar{\mathbf{U}} &= \{\tilde{u}(\bar{x}^n),\tilde{u}'(\bar{x}^n),\tilde{u}''(\bar{x}^n),\ldots\} ,\\ \bar{\mathbf{V}} &= \{\tilde{v}(\bar{x}^n),\tilde{v}'(\bar{x}^n),\tilde{v}''(\bar{x}^n),\ldots\} ,\end{aligned}$$

(' denotes differentiation with respect to ξ). We then introduce a new scalar control variable v by means of the substitution

$$(17.12)\qquad u = \hat{u}(x,v) = \tilde{u}(x^n) + v\tilde{v}(x^n) .$$

x^n of course has now to be regarded as a variable, not as a fixed number. It is then easy to see that the sequence $\mathbf{U}(x,\mathbf{V}) = \{u_0(x,\mathbf{V}),u_1(x,\mathbf{V}),\ldots\}$ which is associated with (17.12) according to (15.4) has this property

$$u_i(x,\mathbf{V}) = \tilde{u}^{(i)}(x^n)+v_0\tilde{v}^{(i)}(x^n) \text{ if } \mathbf{V}= \{v_0,0,0,\ldots\},i=0,1,\ldots$$

($\tilde{u}^{(i)},\tilde{v}^{(i)}$ is the i-th derivative of the polynomial $\tilde{u},\tilde{v}$). The symbol $\mathbf{V}$ for the moment has been used in the same sense as in Sec. 15 and should not be confused with the given fixed $\bar{\mathbf{V}}$. It follows then from (17.11) that for every $\sigma \in \mathbf{R}$

$$(17.13)\qquad \mathbf{U}(\bar{x},\sigma\mathbf{e}) = \bar{\mathbf{U}} + \sigma\bar{\mathbf{V}} \quad\text{where } \mathbf{e} = \{1,0,0,\ldots\} .$$

Next we wish to make use of Theorem 15.2. Let $C,\hat{C}$ respectively be the formal power series associated with the given and the transformed system (i.e. the system $\dot{x} = f(x;\hat{u}(x,v))$) respectively. Taking the particular form of the transformation into account we can apply the simplified version of the theorem (Corollary 2):

*) Note however that $\bar{\mathbf{V}}$ is arbitrary and therefore - in the appropriate context - may be regarded as an indeterminate (e.g.(17.14)!)

$\hat{C}(x,z_1,z_2;\mathbf{V}_0,\mathbf{V}_1,\mathbf{V}_2) =$
$= C(x,z_1,z_2;\mathbf{U}(x,\mathbf{V}_0),\mathbf{U}(x,\mathbf{V}_1),\mathbf{U}(x,\mathbf{V}_2))$.

We now specialize the arguments as follows : $x \to \bar{x}$, $\mathbf{V}_0, \mathbf{V}_2 \to \mathbf{O} = \{0,0,\ldots\}$, $\mathbf{V}_1 \to \sigma\mathbf{e}$, where $\mathbf{e}$ is given as in (17.13) and σ is a parameter. This leads to the following identity in σ, z_1, z_2:

$$\hat{C}(\bar{x},z_1,z_2;\mathbf{O},\sigma\mathbf{e},\mathbf{O}) = C(\bar{x},z_1,z_2;\bar{\mathbf{U}},\bar{\mathbf{U}}+\sigma\bar{\mathbf{V}},\bar{\mathbf{U}}) .$$

Finally if z_i,σ are further specialized according to

$$\sigma = \lambda^r,\ z_1 = \lambda,\ z_2 = \lambda+\lambda^r$$

we obtain both from C and from $\hat{C}$ a power series in λ to which Lemma 17.2 can be applied. We choose $r > \nu$ and compare the coefficient of $\lambda^{3r+\nu-1}$ on both sides. Using the corollary of Lemma 16.2 we arrive then at this result

(17.14) $\hat{c}+\frac{1}{(\nu-1)!}(\hat{\Gamma}^{(\nu-1)}(\partial\hat{B}_0/\partial v_0))(\bar{x},\mathbf{O}) = c+g_\nu(\bar{x},\bar{\mathbf{U}},\bar{\mathbf{V}})$,

where $\hat{c},c$ respectively are linear combinations of $\hat{l}^{(\mu)}(\mathbf{O},\mathbf{e})$ $l^{(\mu)}(\bar{\mathbf{U}},\bar{\mathbf{V}})$ respectively, the coefficients being integers. It is clear therefore that c depends linearly upon the elements of $\bar{\mathbf{V}}$. Next we wish to convince ourselves that $\hat{c}$ also is linear in $\bar{\mathbf{V}}$. To this purpose let us consider the vectors $\hat{B}_\nu(x,\mathbf{V})$ which are associated according to (13.7) with the transformed system $\dot{x} = \hat{f}(x;v) = f(x;\hat{u}(x,v))$, where $\hat{u}$ is given by (17.12) (here again the symbol $\mathbf{V}$ is used in the same sense as in Sec. 15). For every $\nu \geq 0$ we have a relation of the form

(17.15) $\hat{B}_\nu(x,\mathbf{V}) = \sum_{\mu=0}^{\nu}\sum_{\rho=1}^{m} \alpha_{\mu,\rho} B^\rho_\mu(x,\mathbf{U}(x,\mathbf{V}))$,

where $\alpha_{\mu,\rho}$ is a linear form in the derivatives of $\tilde{v}(x^n)$. This is clear for $\nu=0$ since $B^\rho_0 = \partial f/\partial u^{(\rho)}$ and

(17.16) $\hat{B}_0(x,\mathbf{V}) = \partial f(x;\hat{u}(x,v_0)/\partial v_0 = \sum_{\rho=1}^{m} \tilde{v}^{(\rho)}(x^n)(\partial f/\partial u^{(\rho)})(x;\hat{u}(x,v_0))$.

(cf. (17.12). In case of an arbitrary ν it can be inferred from Corollary 2 of Theorem 15.1 and from part (iii) of Lemma 13.1. Note that $\hat{f}^n$ (= n-th component of $\hat{f}$) is equal to 1, hence if $\alpha = \alpha(x^n)$ is a function of x^n alone then we have simply $\hat{\Gamma}(\alpha g) = \alpha\hat{\Gamma}(g)+\beta g$, where $\beta = d\alpha/dx^n$.

It follows now from (17.11), (17.13) and (17.15) that $\hat{B}_\nu(\bar{x},\mathbf{O})$ and hence also $\hat{1}^{(\mu)}(\mathbf{O},\mathbf{e})$ is linear in $\bar{\mathbf{V}}$.

We return to (17.14) and observe that all what remains to be done is to show that for $x=\bar{x}, \mathbf{U}=\bar{\mathbf{U}}, \mathbf{V}=\bar{\mathbf{V}}$ the expression (17.7) coincides with $\hat{\Gamma}^{(\nu-1)}(\partial\hat{B}_0/\partial v_0)(\bar{x},\mathbf{O})$. We proceed in two steps. First we note, that the following relation

$$(\hat{\Gamma}^{(\nu)}(\partial\hat{B}_0/\partial v_0))(x,\mathbf{V}) = (\Gamma^{(\nu)}(\tilde{B}))(x,\mathbf{U})\Big|_{\mathbf{U}\to\mathbf{U}(x,\mathbf{V})}$$

holds for $\nu=0,1,\ldots,$ where

$$\tilde{B}(x,\mathbf{U}) = \sum_{\sigma,\rho=1}^{m} \tilde{v}^{(\rho)}(x^n)\tilde{v}^{(\sigma)}(x^n) \frac{\partial^2 f}{\partial u^{(\rho)}\partial u^{(\sigma)}}(x;u_0) .$$

This relation is an immediate consequence of (17.16) if $\nu=0$. It can be proved for $\nu > 0$ again by combining Corollary 2 of Theorem 15.1 with part (iii) of Lemma 13.1 and Lemma 13.3.

The second step is then obvious. We have to show that $\Gamma^{(\nu-1)}(\tilde{B})$ turns into the expression (17.7) for $x=\bar{x}$, $\mathbf{U}=\bar{\mathbf{U}}$. This however is an easy task in view of (13.2), if one makes use of the relations (17.7), (17.8), (17.11) and takes into account that

$$\frac{\partial^2 f}{\partial u^{(\rho)}\partial u^{(\sigma)}} = \frac{\partial B_0^\rho}{\partial u^{(\sigma)}} .$$

The proof of our next theorem will be based on an evaluation of the formula (17.6)in case the argument $\mathbf{V}$ is replaced by a sequence $\mathbf{V}_k = \{v_{k,0},v_{k,1},\ldots\}$ depending upon an integral valued parameter k. The definition of $\mathbf{V}_k$ is as follows, $k=0,1,\ldots$

$$(17.17) \qquad v_{k,j} = \begin{cases} 0 \text{ for } & j < k , \\ \frac{j!}{(j-k)!} v_{j-k} & \text{for } j \geq k . \end{cases}$$

$\{v_o, v_1, \ldots\} = \mathbf{V}$ is a given sequence of m-dimensional vectors, hence $\mathbf{V}_k$ is actually a function of k and $\mathbf{V}$. In the course of our next considerations the argument x and occasionally also $\mathbf{U}$ will be omitted.

Lemma 17.3. The following relations hold
(i) $l^{(\nu)}(\mathbf{U}, \mathbf{V}_k) = 0$ if $\nu \leq k$. (ii) $l^{(k+1+\sigma)}(\mathbf{U}, \mathbf{V}_k) = ((k+\sigma)!)(\sigma!)^{-1} l^{(\sigma+1)}(\mathbf{U}, \mathbf{V})$ if $\sigma \geq 0$. (iii) If $\tau \geq 0$ then

$$\frac{1}{(2k+\tau)!} L^{(2k+1+\tau)}(\mathbf{U}, \mathbf{V}_k) \cdot \mathbf{V}_k =$$

$$= \sum_{\rho,\sigma=1}^{m} \sum_{i+j+\omega \leq \tau} \chi(i,j,\omega;k) v_i^{(\sigma)} v_j^{(\rho)} (\Gamma^{\omega}(\partial B^{\rho}_{\tau-i-j-\omega}/\partial u_o^{(\sigma)}))(\mathbf{U}) ,$$

where

$$\chi(i,j,\omega;k) = (k+j+\omega-1)!(j+k)((k+\tau-i)!)^{-1}(i!j!\omega!)^{-1} .$$

It should be noted that the restriction $i+j+\omega \leq \tau$ implies that χ can also be written in this form

$$(17.18) \qquad \chi(i,j,\omega;k) = (j+k)(i!j!\omega!)^{-1} \prod_{\varphi=j+\omega}^{\tau-i} (k+\varphi)^{-1} .$$

Proof. (i) follows immediately from (17.17) and Theorem 13.1. From the same theorem we obtain

$$l^{(1+\sigma+k)}(\mathbf{U}, \mathbf{V}_k) = \sum_{\rho=k}^{k+\sigma} \binom{k+\sigma}{\rho} B_{k+\sigma-\rho} \frac{\rho!}{(\rho-k)!} v_{\rho-k}$$

$$= \frac{(k+\sigma)!}{\sigma!} \sum_{\rho=k}^{k+\sigma} \frac{\sigma!}{(\rho-k)!(\sigma-\rho+k)!} B_{\sigma-\rho+k} v_{\rho-k}$$

$$= ((k+\sigma)!)(\sigma!)^{-1} l^{(\sigma+1)}(\mathbf{U}, \mathbf{V}) .$$

In order to prove (iii) one has to go back to Definition 12.1

$$\frac{1}{(\nu-1)!}L^{(\nu)}(\mathbf{U},\mathbf{V})\cdot\mathbf{V}=\sum_{i=0}^{\nu-1}\sum_{j=0}^{\nu-1}\frac{1}{(\nu-1-i)!i!}\left(\frac{\partial B_{\nu-1-i}\cdot v_i}{\partial u_j}\right)\cdot v_j \quad .$$

Here and in the sequel we adopt the abbreviation $(\partial(B_{\nu-1-i}\cdot v_i)/\partial u_j)\cdot v_j$ in order to denote the vector

$$\sum_{\rho,\sigma=1}^{m} v_i^{(\sigma)} v_j^{(\rho)}\,\partial B^{\sigma}_{\nu-1-i}/\partial u_j^{(\rho)} \quad ,$$

(Note that $v_i^{(\sigma)}$ are components of v_i) . **We will** assume **in the sequel** that $v_0=0$. Using the identity (13.10) and the linearity of the operator Γ we can rewrite the expression on the right hand side of the above formula as follows

$$\sum_{i=0}^{\nu-1}\sum_{j=0}^{\nu-1}\frac{1}{(\nu-1-i)!i!}\sum_{\omega}\binom{j+\omega-1}{\omega}\Gamma^{\omega}\left(\frac{\partial B_{\nu-1-i-j-\omega}\cdot v_i}{\partial u_0}\right)\cdot v_j =$$

(17.19)

$$=\sum_{i=0}^{\nu-1}\sum_{j=0}^{\nu-1}\sum_{\omega}\frac{(j+\omega-1)!j}{(\nu-1-i)!\omega!}\;\frac{1}{i!j!}\;\Gamma^{\omega}\left(\frac{\partial B_{\nu-1-i-j-\omega}\cdot v_i}{\partial u_0}\right)\cdot v_j \quad .$$

The sum over ω has to be extended from 0 to $\nu-1-i-j$. We now substitute $\mathbf{V}\to\mathbf{V}_k$, $\nu\to 2k+\tau+1$. The terms in the above sum which possibly remain different from zero are then those for which

$$k\le i,j \qquad , \quad 2k+\tau-i-j-\omega\ge 0 \ .$$

The statement of the lemma follows now simply by using (17.17) and changing from i,j to $i-k$ and $j-k$ as variables of summation.

<u>Theorem 17.2</u> The following identity in $\mathbf{t},\mathbf{x},\mathbf{U}$ holds

$$[B^{\rho}_{\mu},B^{\sigma}_{\nu}] = (-1)^{\mu}\sum_{\tau=0}^{\mu+1}\binom{\mu+1}{\tau}(-1)^{\tau}\Gamma^{\tau}((\partial B^{\rho}_{\nu+\mu+1-\tau}/\partial u_0^{(\sigma)})) \ .$$

<u>Proof.</u> The proof **is given first in the autonomous case.**

<u>Step 1.</u> We claim that $g_{2k+\tau+1}(\mathbf{U},\mathbf{V}_k)$ is independent from k , where $g_\nu(\mathbf{U},\mathbf{V})$ is the function given by (17.2) and $\mathbf{V}_k$ the sequence defined according to (17.17). The most convenient way to

verify this statement is to use the representation (17.7) for $(\nu-1)!g_\nu$ (cf. Theorem 17.1). It follows from the Leibniz-rule that $\pi^{(\lambda)}_{\sigma,\rho}(\mathbf{V}_k) = 0$ if $\nu-1-\lambda < 2k$ i.e. if $\lambda > \nu-1-2k$. Furthermore one sees from (17.8) that

$$\pi^{(\lambda)}_{\sigma,\rho}(\mathbf{V}_k) = \sum_{\mu+\xi=2k+\tau-\lambda} \frac{(2k+\tau-\lambda)!}{\mu!\xi!} v^{(\sigma)}_{k,\mu} v^{(\rho)}_{k,\xi}$$

in case $\nu=2k+\tau+1$. Using (17.17) we finally obtain for $\nu=2k+\tau+1$

$$\pi^{(\lambda)}_{\sigma,\rho}(\mathbf{V}_k) = \begin{cases} 0 \text{ if } \lambda > \tau , \\ (2k+\tau-\lambda)! \sum_{\mu+\xi=\tau-\lambda} \frac{1}{\mu!\xi!} v^{(\sigma)}_{\mu} v^{(\rho)}_{\xi} \text{ if } \lambda \le \tau . \end{cases}$$

Hence for $\nu = 2k+\tau+1$ and $\mathbf{V} = \mathbf{V}_k$ all terms in (17.7) vanish except those for which $\lambda \le \tau$. The latter ones can be written in the form $(2k+\tau)!$ times a vector which is independent from k .

Step 2. It follows from Lemma 17.3 and from (17.2) that we have

$$g_{2k+\tau+1}(\mathbf{U}, \mathbf{V}_k) = \frac{1}{(2k+\tau)!} L^{(2k+\tau+1)}(\mathbf{U}, \mathbf{V}_k)\cdot \mathbf{V}_k +$$

$$+ \sum_{\substack{\lambda+\mu=\tau+1 \\ \lambda>0,\mu>0}} \frac{1}{(\lambda-1)!(\mu-1)!(k+\mu)} [1^{(\lambda)}(\mathbf{U},\mathbf{V}), 1^{(\mu)}(\mathbf{U},\mathbf{V})] .$$

On the other hand we know from the first part of the proof that the expression on the right hand side of the above formula remains unchanged if the paramter k varies on the positive integers. Using part (iii) of Lemma 17.3 we conclude by standard arguments that for any integer $\tau \ge 0$ the following relation holds identically in ξ

$$\sum_{\substack{\lambda+\mu=\tau+1 \\ \lambda>0,\mu>0}} \frac{1}{(\lambda-1)!(\mu-1)!(\xi+\mu)} [1^{(\mu)}(\mathbf{U},\mathbf{V}), 1^{(\lambda)}(\mathbf{U},\mathbf{V})] =$$

$$= \sum_{\rho,\sigma=1}^{m} \sum_{i+j+\omega\le\tau} \chi(i,j,\omega;\xi) v^{(\sigma)}_i v^{(\rho)}_j (\Gamma^{\omega}(\partial B^{\rho}_{\tau-i-j-\omega}/\partial u^{(\sigma)}_0))(\mathbf{U})$$

plus a vector which is independent from ξ . The symbol ξ now represents a scalar parameter which is allowed to vary in the whole real field $\mathbb{R}$. χ is the rational function given by

(17.20) $\chi(i,j,\omega;\xi) = (\xi+j)(i!j!\omega!)^{-1} \prod_{\varphi=j+\omega}^{\xi-j} (\xi+\varphi)^{-1}$

(cf. (17.18)). If we multiply the above relation by $\xi+\mu$ and let then $\xi \to -\mu$ we obtain finally

$$((\lambda-1)!(\mu-1)!)^{-1}[1^{(\mu)}(\mathbf{U},\mathbf{V}),1^{(\lambda)}(\mathbf{U},\mathbf{V})] =$$

(17.21)
$$= \sum_{\rho,\sigma=1}^{m} \sum_{i+j+\omega\leq\tau} \Psi(i,j,\omega,\mu)v_i^{(\sigma)}v_j^{(\rho)}\Gamma^{\omega}(\partial B^{\rho}_{\tau-i-j-\omega}/\partial u_o^{(\sigma)}))(\mathbf{U}),$$

where $\tau = \lambda+\mu-1$ and

$$\Psi(i,j,\omega,\mu)=\lim_{\xi\to-\mu} (\xi+\mu)\chi(i,j,\omega;\xi) .$$

Using (17.20) Ψ is easily computed:

(17.22) $\Psi(i,j,\omega,\mu)=\frac{(j-\mu)(-1)^{\mu-j-\omega}}{i!j!\omega!(\mu-j-\omega)!(\lambda-1-i)!}$ if $i \leq \lambda-1$, $0 \leq \omega \leq \mu-j$

and $\Psi(\ldots)=0$ otherwise. The proof is now concluded by using the representation of $1^{(\nu)}$ given in Theorem 13.1:

$$((\nu-1)!)^{-1}1^{(\nu)}(\mathbf{U},\mathbf{V}) = \sum_{\rho=1}^{m} \sum_{j=0}^{\nu-1} \frac{1}{j!(\nu-1-j)!} B^{\rho}_{\nu-1-j} v_j^{(\rho)} .$$

One simply has to compare the coefficients of a fixed term $v_i^{(\sigma)}v_j^{(\rho)}$ on both sides of (17.21). Let us assume that

(17.23) $j < \mu$ and $i < \lambda$, but not $i < \mu$ and $j < \lambda$.

Then the coefficient on the left hand side is equal to

(17.24) $\frac{1}{i!j!(\mu-1-j)!(\lambda-1-i)!}[B^{\rho}_{\mu-1-j} , B^{\sigma}_{\lambda-1-i}]$

and on the right hand side

(17.25) $\sum_{0\leq\omega\leq\mu-j} \frac{(j-\mu)(-1)^{\mu-j-\omega}}{i!j!\omega!(\mu-j-\omega)!(\lambda-1-i)!} \Gamma^{\omega}(\partial B^{\rho}_{\tau-i-j-\omega}/\partial u_o^{(\sigma)})$

where $\tau=\lambda+\mu-1$ (cf. (17.22)) . Given now two positive integers r,s. We choose two further positive integers i,j such that $s < i-j$ and put

$$\lambda = r+i, \quad \mu = s+j < i .$$

It is then clear that (17.23) is satisfied, and we can therefore for this choice of λ,μ,i,j identify the expressions (17.24) and (17.25). This yields the following result

$$\frac{1}{(s-1)!(r-1)!}[B^{\rho}_{s-1}, B^{\sigma}_{r-1}] =$$

$$= -\sum_{\omega=0}^{s} (-1)^{s-\omega} \frac{s}{\omega!(s-\omega)!(r-1)!} \Gamma^{\omega}(\partial B^{\rho}_{r+s-1-\omega}/\partial u_o^{(\sigma)})$$

which can, by an obvious change of notation, given the form as was stated in the theorem.

Thereby the theorem is proved for those control systems where the right hand side does not depend explicitly upon t. In order to remove this proviso one simply can apply standard space augmentation technique, i.e. consider $x^{n+1} := t$ as additional state variable. We put $x^* := (x^T, x^{n+1})^T$, $f^*(x^*;u) := (f(x^{n+1},x;u)^T,1)^T$. Accordingly we denote by $\Gamma^*, B^*_{\nu}(x^*,\mathsf{U})$ respectively the operator and the sequence of (n+1)-dimensional vectors respectively associated with the augmented system

(17.26) $\quad \dot{x}^* = f^*(x^*;u).$

Clearly

(17.27) $\quad B^*_o(x^*,\mathsf{U}) = (\partial f^*/\partial u)(x^*;u) = (B_o(x^{n+1},x,u)^T,0)^T$.

We are then in the position to write down the statement of the theorem with $B^*_{\nu} := (\Gamma^*)^{\nu} B^*_o$ instead of B_{ν} and $(\Gamma^*)^{\nu}$ instead of Γ^{ν}. The desired result is now obtained by just comparing the n first components on the left and right hand side and writing t instead of x^{n+1}. All this follows immediately from the subsequent lemma

<u>Lemma 17.3.</u> Let $g(t,x,\mathsf{U}), h(t,x,\mathsf{U})$ be n-dimensional vectors depending upon t,x,U and let

(17.28) $\quad g^*(x^*,\mathsf{U}) := (g(x^{n+1},x,\mathsf{U})^T,0)^T, \; h^*(x^*,\mathsf{U}) := (h(x^{n+1},x,\mathsf{U})^T,0)^T$

be the corresponding augmented vectors. Claim:

(i) $[g^*,h^*](x^*,\mathcal{U})\Big|_{x^{n+1}\to t} = (([g,h](t,x,\mathcal{U}))^T,0)^T$

where the Lie-bracket on the left and right hand side respectively has to be performed with respect to the state variable x^* and x respectively,

(ii) $(\Gamma^*)^\nu(g^*)(x^*,\mathcal{U})\Big|_{x^{n+1}\to t} = (([\Gamma^\nu(g)(t,x,\mathcal{U}))^T,0)^T.$

Proof. (i) follows by inspection from (17.28). (ii) can be rewritten in this form

$$(\Gamma^*)^\nu(g^*)(x^*,\mathcal{U}) = ((\Gamma^\nu(g)(t,x,\mathcal{U}))^T,0)^T \quad \text{with} \quad t \to x^{n+1},$$

It is then obvious that the last relation holds for arbitrary $\nu > 0$ if it holds for $\nu = 1$. Hence it is sufficient to prove (ii) in case $\nu=1$ and this again can be handled by inspection (arguments are omitted):

$$\Gamma^*(g^*) = \begin{pmatrix} \sum_i (\partial g/\partial u^i)u^{i+1} \\ 0 \end{pmatrix} + \begin{pmatrix} g_x & g_t \\ 0 & 0 \end{pmatrix}\begin{pmatrix} f \\ 1 \end{pmatrix} - \begin{pmatrix} f_x & f_t \\ 0 & 0 \end{pmatrix}\begin{pmatrix} g \\ 0 \end{pmatrix}$$

$$= \begin{pmatrix} \Gamma(g) \\ 0 \end{pmatrix}.$$

18. A special case of the Clebsch-Legendre condition.

The generalized Clebsch-Legendre condition will be established in full generality in the next sections. In this section we deal with a special case; it will turn out that the proof of the general case can then be reduced without much difficulty to the situation considered in Lemma 18.3.

We begin with a lemma which is of more technical nature and is needed subsequently.

Lemma 18.1. Given positive integers μ,τ. Then there exists a positive integer $\varkappa_0$ such that for every $\varkappa \geq \varkappa_0$ one can find real numbers ξ_i, z_i, $i=1,\ldots,\tau+3$ having the following properties

$$(i) \quad \sum_{i=1}^{\tau+3} (\xi_{i-1}-\xi_i) z_i^{\nu} = 0 \quad \text{for} \quad \nu = \varkappa+1, \ldots, \varkappa+\tau, \; \xi_0 = \xi_{\tau+3} = 0$$

$$(18.1)(ii) \quad z_1 < z_2 < \ldots < z_{\tau+3} < 0 ,$$

(iii) $\zeta \neq 0$, where ζ is given as

$$\frac{1}{2\varkappa+1+\mu}\Big(\prod_{\lambda=1}^{\mu} \frac{1}{\varkappa+\lambda}\Big) \sum_{i=1}^{\tau+3} (\xi_{i-1}-\xi_i)(\xi_{i-1}+\xi_i) z_i^{2\varkappa+1+\mu} -$$

$$- \sum_{1 \le i < j \le \tau+3} \; \sum_{\sigma+\rho=\mu-1} \frac{(-1)^{\sigma}}{\sigma!\rho!(\sigma+1+\varkappa)(\rho+1+\varkappa)} (\xi_{j-1}-\xi_j)(\xi_{i-1}-\xi_i) z_j^{\sigma+1+\varkappa} z_i^{\rho+1+\varkappa} .$$

Proof. We begin with the following observation. If, for a certain $\varkappa > 0$, one can find ξ_i, z_i such that (i), (iii) and

$$(18.2) \quad z_i \neq 0, \quad z_i \neq z_j \quad \text{for} \quad i \neq j , \; i,j \le \tau+1; \; z_{\tau+3} = 0$$

holds, then one can also determine ξ_i, z_i such that all three conditions (i) - (iii) hold. Indeed, if we regard $\xi_{\tau+2}, \xi_{\tau+1}$ as fixed and solve the equations (i) formally with respect to ξ_i , $i=1,\ldots,\tau$, we obtain these quantities and hence also $\xi_{\tau+3} = \xi_0 = 0, \xi_1, \ldots, \xi_{\tau+2}$ as rational functions of the z_i, the denominator Δ being a rational function of the z_i which does not vanish on (18.2) . Inserting these functions into the expression for ζ we obtain a rational function of the z_i with Δ^2 as denominator and with a numerator which cannot vanish identically in $z_1, \ldots, z_{\tau+3}$. Otherwise one could not specialize the z_i such that (18.2) holds and ζ assumes a value $\neq 0$. Hence, by standard arguments the numerator can also not vanish on the set of all $(z_1, \ldots, z_{\tau+3})$ which is defined in terms of the inequalities (18.1) (ii) . So we now prove the following modified version of the lemma: There exist for $\varkappa \ge \varkappa_0$, real numbers ξ_i, z_i, $i=1,\ldots, \tau+3$, satisfying conditions (18.1), (i) and (iii) and (18.2).

We choose fixed positive z_i such that (18.2) holds and such that we have in addition for all $i,j \le \tau+2$

(18.3) $z_i/z_j \neq z_{i'}/z_{j'}$ if $i > j$, $i' > j'$ and $(i,j) \neq (i',j')$.

Furthermore we put

$$\xi_i = \sum_{j=1}^{i} \eta_j z_j^{-\varkappa} \ , \ i=0,\dots,\tau+2 \ .$$

We have then, for $1 \le i \le \tau+2$

$$\xi_i - \xi_{i-1} = \eta_i z_i^{-\varkappa} \ , \ \xi_i + \xi_{i-1} = \Big(\eta_i + 2\sum_{j=1}^{i-1} \eta_j \Big(\frac{z_i}{z_j}\Big)^{\varkappa}\Big) z_i^{-\varkappa} \ ,$$

and the conditions (18.1), (i) and (iii) can be written in the form

(18.4) $$\sum_{i=1}^{\tau+2} \eta_i z_i^{\nu} = 0 \quad \text{for } \nu=1,\dots,\tau \ ,$$

(18.5) $$\sum_{i=1}^{\tau+2} \eta_i \Big(\eta_i + 2\sum_{j=1}^{i-1} \eta_j \Big(\frac{z_i}{z_j}\Big)^{\varkappa}\Big) z_i^{1+\mu} + \omega(\varkappa) \neq 0 \ ,$$

where $\omega(\varkappa)$ is a certain rational function of $\varkappa$. We now choose $\eta_{\tau+1} \neq 0$, $\eta_{\tau+2} \neq 0$ and determine the remaining η_i such that equations(18.4) are satisfied. The expression (18.5) can then be written in the form

(18.6) $$\sum_{1 \le j < i \le \tau+2} \rho_{i,j} \Big(\frac{z_i}{z_j}\Big)^{\varkappa} + \omega(\varkappa)$$

with $\rho_{\tau+2,\tau+1} = 2\eta_{\tau+2}\, \eta_{\tau+1}\, z_{\tau+2}^{1+\mu} \neq 0$.

It is then clear, in view of (18.3), that (18.6) will be different from 0 for all $\varkappa \ge \varkappa_0$, $\varkappa_0$ being some sufficiently large number.

Next we introduce a special class of power series in a scalar variable λ which are of the type (8.18). Combining the results of Sec. 14 and Sec. 17 we are now in a position to analyse the coefficients which are relevant for the second order conditions. t,x in the sequel is mostly ommitted or indicated by an asterisk.

We first calculate $l^{(\nu)}$, $L^{(\nu)}$ for special values of the arguments. Let ξ be a scalar variable and $\varkappa$ a non-negative integer. We denote by $\mathbf{V}(\xi,\varkappa)$ the sequence $\{v_0(\xi,\varkappa),v_1(\xi,\varkappa),\ldots\}$ of m-dimensional vectors having this property

(18.7) $v_j(\xi,\varkappa) = 0$ if $j \neq \varkappa$, $v_\varkappa(\xi,\varkappa) = (\xi,0,\ldots,0)^T$

(i.e. $v_\varkappa$ is the m-tuple having ξ as first component and 0 elsewhere).

Assume now that there is given a sequence $\mathbf{U} = \{u_0,u_1,\ldots.\}$, positive integers $\pi,\varkappa$ and real numbers $\xi_1,\ldots,\xi_M$. We put $\xi_0 = 0$. Let λ be a real parameter and

(18.8) $\mathbf{U}_0 = \mathbf{U}$, $\mathbf{U}_i = \mathbf{U} + \mathbf{V}(\lambda^\pi \xi_i,\varkappa)$ $i=1,\ldots,M$.

We have then, in view of (18.7)

$$l^{(\nu)}(\mathbf{U}_0, \mathbf{U}_{i-1} - \mathbf{U}_i) = \begin{cases} 0 & \text{if } \nu \leq \varkappa \;, \\ \lambda^\pi \binom{\nu-1}{\varkappa}(\xi_{i-1}-\xi_i)B^1_{\nu-1-\varkappa} & \text{if } \nu > \varkappa \;. \end{cases}$$

$$L^{(\nu)}(\mathbf{U}_0, \mathbf{U}_{i-1} - \mathbf{U}_i)\cdot(\mathbf{U}_{i-1} + \mathbf{U}_i - 2\mathbf{U}_0) = \begin{cases} 0 & \text{if } \nu \leq \varkappa \\ \lambda^{2\pi}\binom{\nu-1}{\varkappa}(\xi_{i-1}-\xi_i)(\xi_{i-1}+\xi_i)\dfrac{\partial B^1_{\nu-1-\varkappa}}{\partial u^{(1)}_\varkappa} & \\ \quad \text{if } \nu > \varkappa \;, \end{cases}$$

(Note that B^1_ν is the first column of the matrix B_ν).

<u>Lemma 18.2.</u> Let $\mathbf{V}$, $\mathbf{U}_i$, $i=1,\ldots,M$ be defined according to (18.7), (18.8) and let the symbol $\tilde{C}$ denote the formal series $C(x,\lambda z_1,\ldots,\lambda z_M, \mathbf{U}_0, \mathbf{U}_1,\ldots,\mathbf{U}_M)$. Similarly let $\tilde{C}^{(1)}$, $\tilde{C}^{(2)}$, $\tilde{\check{C}}^{(2)}$, $\tilde{R}_{\sigma,\rho}$ be the quantities derived from the respective quantities defined in Sec. 14 by means of the same specializations of the arguments z_i and $\mathbf{U}_i$. Then the following relations hold

(i) $\tilde{C} = \tilde{C}^{(2)} + \mathcal{O}(\lambda^{3\pi+1})$

(ii) $\tilde{R}_{\sigma,\rho}=0$ if $\sigma \le \varkappa$ or $\rho \le \varkappa$,

$$\tilde{R}_{\sigma,\rho}=-\lambda^{2\pi+\sigma+\rho}\Big\{\sum_{1\le i<j\le M}\binom{\sigma-1}{\varkappa}\binom{\rho-1}{\varkappa}(\xi_{j-1}-\xi_j)(\xi_{i-1}-\xi_i)z_j^\sigma z_i^\rho[B^1_{\sigma-1-\varkappa},B^1_{\rho-1-\varkappa}]$$

$$+\Big(\sum_{i=1}^{M}(\xi_{i-1}-\xi_i)z_i^\sigma\Big)\Big(\sum_{i=1}^{M}(\xi_{i-1}-\xi_i)z_i^\rho\Big)\tilde{R}'_{\sigma,\rho}\Big\} \quad \sigma>\varkappa \text{ and } \rho>\varkappa ,$$

where $\tilde{R}'_{\sigma,\rho}=\binom{\rho-1}{\varkappa}\binom{\sigma-1}{\varkappa}(B^1_{\rho-1-\varkappa})_x B^1_{\sigma-1-\varkappa}$.

$$\text{(iii)}\ \tilde{C}^{(2)}=\tilde{C}^{(1)}+\tfrac{1}{2}\lambda^{2\pi}\sum_{\nu=\varkappa+1}^{\infty}\frac{\lambda^\nu}{\nu!}\binom{\nu-1}{\varkappa}\Big(\sum_{i=1}^{M}(\xi_{i-1}-\xi_i)(\xi_{i-1}+\xi_i)z_i^\nu\ \frac{\partial B^1_{\nu-1-\varkappa}}{\partial u^{(1)}_\varkappa}\Big)+\check{\tilde{C}}^{(2)}$$

$$\text{(iv)}\ \tilde{C}^{(1)}=x+\lambda^{\pi}\sum_{\nu=\varkappa+1}^{\infty}\frac{\lambda^\nu}{\nu!}\binom{\nu-1}{\varkappa}\Big(\sum_{i=1}^{M}(\xi_{i-1}-\xi_i)z_i^\nu\Big)B^1_{\nu-1-\varkappa} \quad .$$

Proof. (i) follows from the definition of $C^{(2)}$ (cf. Sec. 14) and from the fact that $\mathcal{U}_i-\mathcal{U}_o=\mathcal{O}(\lambda^\pi)$. (cf. (18.8)). The remaining relations are an immediate consequence of (14.3), (14.6) and the formulas for $l^{(\nu)}$, $L^{(\nu)}$ which were derived from (18.8) .

Lemma 18.3. Given x, $\mathcal{U}=\{u_o,u_1,\dots\}$ and assume that (8.19) is satisfied. We denote by $\mathcal{L}$ be the linear subspace of $\mathbb{R}^n$ which is spanned by the columns B^i_ν of the matrices $B_\nu=B_\nu(t,x,\mathcal{U})$, $\nu=0,1,\dots,i=1,\dots,m$. Let furthermore $\mu \ge 0$ be an even integer and let us assume that the following n-dimensional vectors (having t x, $\mathcal{U}$ as its argument) are contained in $\mathcal{L}$:

(18.9)
(i) $\partial B^1_{\sigma+\nu}/\partial u^{(1)}_\sigma$ for $\sigma=0,1,\dots$ and $\nu<\mu$,
(ii) $\partial B^1_{\sigma+\mu}/\partial u^{(1)}_\sigma - \partial B^1_\mu/\partial u^{(1)}_o$ for $\sigma=0,1,\dots$,
(iii) $[B^1_\sigma,B^1_\rho]$ if $\sigma+\rho<\mu-1$,
(iv) $[B^1_\sigma,B^1_\rho]-(-1)^\sigma\partial B^1_\mu/\partial u^{(1)}_o$ if $\sigma+\rho=\mu-1$.

Claim: $(-1)^{\mu/2}\partial B^1_\mu/\partial u_o^{(1)} \in \mathbf{P}_U(t,x,U)$

<u>Proof.</u> Let d be a positive integer such that $\mathcal{L}$ is spanned by the columns of the matrices B_ν, $\nu < d$. We choose a positive integer τ such that

(18.10) $\mu + 1 + 2d \leq 2\tau \quad , \quad 2\mu + 1 \leq 2\tau$.

We will now actually establish the following result:

(18.11) $\quad \zeta(\mu,\tau)\partial B^1_\mu/\partial u_o^{(1)} \in \mathbf{P}_U(t,x,U)$.

where $\zeta(\mu,\tau) \neq 0$ is a certain number depending upon μ,τ (but not upon the underlying system (7.1) and in particular not upon the dimension m of u). In the next section the sign of this number will then be determined for any pair μ,τ, where $\mu \geq 0$ and $2\mu+1 \leq 2\tau$, and it will turn out that the sign is equal to $(-1)^{\mu/2}$.

Having fixed μ,τ we construct real numbers ξ_i, z_i, $i=1,\ldots,\tau+3=M$ and a positive integer $\varkappa$ such that all conditions (18.1) are satisfied and that we have in addition

(18.12) $\quad 0 < \varkappa+1-\tau+2\mu \quad , \quad 0 < \varkappa+1-\tau+\mu+d$.

This is possible, in view of Lemma 18.1. Finally we choose a positive integer π such that

$$\begin{aligned} \pi + \varkappa + \tau &< 2\pi + 2\varkappa + 1 + \mu \\ \tau &\leq \pi + 1 \\ 2\varkappa + 1 + \mu &\leq \pi . \end{aligned} \tag{18.13}$$

We then put

(18.14) $\quad r = \pi + \varkappa + \tau \quad , \quad s = 2\pi + 2\varkappa + 1 + \mu$,

and remark that we have, as a consequence of the arrangements made so far, these inequalities:

(18.15)
(i) $r \le s$,
(ii) $s \le 2\,(r-d)$,
(iii) $s \le 3\pi$,
(iv) $s < 2\pi + 2\varkappa + 2\tau$,
(v) $r \le 2\pi + \varkappa + 1$.

(i) follows from (18.13) , (ii) from (18.10) , (iii) from the last of the relations (18.13), (iv) from (18.10), (v) from the second of the relations (18.13). ξ_i, z_i, $i=1,\dots,M = \tau+3$ and the integers $\pi,\varkappa$ are from now on regarded as fixed, $\lambda \ge 0$ as variable. $\mathfrak{U}_i$ are defined according to (18.8) and hence functions of λ . We consider

$$\tilde{C}(\lambda) = C(\divideontimes, \lambda z_1, \dots, \lambda z_M, \mathfrak{U}_0, \mathfrak{U}_1, \dots, \mathfrak{U}_M)$$

and wish to apply the second corollary to Lemma 14.2.
To this purpose we first observe that the hypothesis (14.7) of this corollary is satisfied because of (18.15), (i) , (ii). Also the first of the conditions (14.9) holds in the present situation, since $z_i(\lambda) = \lambda z_i$. The second condition holds because of $\xi_{\tau+3} = \xi_M = 0$ and hence $\mathfrak{U}_M = \mathfrak{U}_0$ (cf. (18.8)) . Finally it follows from statement (i) of Lemma 18.2 and from (18.15), part (iii) that

$$\tilde{C}(\lambda) = \tilde{C}^{(2)}(\lambda) + \mathcal{O}(\lambda^{s+1}) .$$

The statement (18.11) is therefore proved - in view of the said corollary - if we have established the following result

$$\tilde{C}^{(2)}(\lambda) = x + \lambda^r \sum_\nu \lambda^\nu b_\nu + \lambda^s p + \mathcal{O}(\lambda^{s+1}) \quad \text{where}$$

(18.16)

$$b_\nu \in \mathcal{L} , \; p = \frac{1}{2}(\varkappa !)^{-2} \zeta\, \partial B_\mu / \partial u_0^{(1)} ,$$

ζ being the real number appearing in Lemma 18.1, condition (iii). This is now easily done by using the hypothesis (18.9) and the fact that $\tau+3=M$ and hence

$$(18.17) \quad \sum_{i=1}^{M} (\xi_{i-1} - \xi_i) z_i^\nu = 0 \quad \text{for } \nu = \varkappa+1, \dots, \varkappa+\tau .$$

(cf. Lemma 18.1). It follows from this relation and from part (ii) of Lemma 18.2 that $\tilde{R}_{\sigma,\rho} = 0$ if $\sigma \le \varkappa$ or $\rho \le \varkappa$, $\tilde{R}_{\sigma,\rho} = \lambda^{2\pi+\sigma+\rho} b_{\sigma,\rho}$ with $b_{\sigma,\rho} \in \mathcal{L}$, if $\sigma > \varkappa$ and $\rho > \varkappa$ but $\sigma+\rho < 2\varkappa+1+\mu$, and $\tilde{R}_{\sigma,\rho} = -\lambda^{s}(b'+\tilde{\zeta}_{\sigma,\rho}\,\partial B^1_\mu/\partial u^{(1)}_0)$ if $\sigma > \varkappa$, $\rho > \varkappa$ and $\sigma+\rho = 2\varkappa+1+\mu$, where $b' \in \mathcal{L}$ and

$$(18.18)\quad \tilde{\zeta}_{\sigma,\rho} = \binom{\sigma-1}{\varkappa}\binom{\rho-1}{\varkappa}(-1)^{\sigma-1-\varkappa}\sum_{1\le i<j\le M}(\xi_{j-1}-\xi_j)(\xi_{i-1}-\xi_i)z_j^{\sigma}z_i^{\rho} .$$

Hence we have, in view of (14.6) and because of $2\pi + \sigma + \rho > r$ if $\sigma > \varkappa$, $\rho > \varkappa$ (cf. (18.15), (v)) :

$$(18.19)\quad \check{C}^{(2)}(\lambda) = \sum_{\nu>r} \lambda^{\nu} b_{\nu} + \lambda^{s}\zeta'\,\partial B^1_\mu/\partial u^{(1)}_0 + \mathcal{O}(\lambda^{s+1})$$

where $b_\nu \in \mathcal{L}$ and

$$(18.20)\quad \zeta' = -\frac{1}{2}\sum_{\substack{\sigma+\rho=2\varkappa+1+\mu\\ \sigma,\rho>\varkappa}} \frac{1}{\sigma!\rho!}\tilde{\zeta}_{\sigma,\rho}, \text{ where } \tilde{\zeta}_{\cdot,\cdot} \text{ is given by (18.18).}$$

Furthermore, since $r = \pi+\varkappa+\tau$ (cf. (18.14)) we obtain from (18.17) and from part (iv) of Lemma 18.2 this relation

$$(18.21)\quad \tilde{C}^{(1)}(\lambda) = x + \lambda^{r}\sum_{\nu}\lambda^{\nu}b_{\nu} \quad , \quad b_\nu \in \mathcal{L}.$$

Finally it follows from (13.8) and from (18.9), (i) and (ii) that

$$\partial B^1_{\nu-1-\varkappa}/\partial u^{(1)}_{\varkappa} \begin{cases} = 0 & \text{if } \nu < 2\varkappa+1 , \\ \in \mathcal{L} & \text{if } \nu < 2\varkappa+1+\mu . \end{cases}$$

and

$$\partial B^1_{\nu-1-\varkappa}/\partial u^{(1)}_{\varkappa} - \partial B^1_{\mu}/\partial u^{(1)}_0 \in \mathcal{L} \text{ if } \nu = 2\varkappa+1+\mu .$$

Hence we have, in view of Lemma 18.2, part (iii), and since $r \le 2\pi + \varkappa + 1$ (cf. (18.15))

$$\tilde{C}^{(2)} = \tilde{C}^{(1)} + \check{C}^{(2)} + \lambda^{r}\sum_{\nu}\lambda^{\nu}b_{\nu} + \lambda^{s}\zeta''\,\partial B^1_\mu/\partial u^{(1)}_0 + \mathcal{O}(\lambda^{s+1})$$

where $b_\nu \in \mathcal{L}$ and

(18.22) $$\varsigma'' = \frac{1}{2}\,\frac{1}{(2\varkappa+1+\mu)!}\binom{2\varkappa+\mu}{\varkappa}\sum_{i=1}^{M}(\xi_{i-1}-\xi_i)(\xi_{i-1}+\xi_i)z_i^{2\varkappa+1+\mu} \quad .$$

It follows then from the last relations and from (18.19), (18.21) that (18.16) indeed holds true with

$$\varsigma = 2(\varkappa!)^2(\varsigma'+\varsigma'') \quad .$$

One easily recognizes from (18.20) and (18.22) this number as the same which appears in Lemma 18.1 (note that $M = \tau+3$) .

19. Auxiliary results.

Lemma 19.1 Given a solution $u(\cdot),x(\cdot)$ of (7.1) on some interval I and let us denote by $\mathbf{U}(t)$ and $\mathcal{L}(t)$ respectively the sequence $\{u(t),\dot{u}(t),\ldots\}$ and the linear space spanned by the columns $B^i_\nu(t,x(t),\mathbf{U}(t))$ of the matrices $B_\nu(t,x(t),\mathbf{U}(t))$, $\nu = 0,1,\ldots,i=1,\ldots,m$, $t \in I$. Assume that the dimension of $\mathcal{L}(t)$ is independent from t and that we have

(19.1) $(\partial B^1_\nu/\partial u^{(1)}_o)(t,x(t),\mathbf{U}(t)) \in \mathcal{L}(t)$

for all $t \in I$ and $\nu < \mu$, μ being some positive integer.

Claim: All the vectors listed under (18.9) and evaluated for t, $x=x(t)$, $\mathbf{U} = \mathbf{U}(t)$ belong to the space $\mathcal{L}(t)$.

Proof. We infer from Corollary 1 to Lemma 13.1 and from (19.1) that

$$(\Gamma^\tau(\partial B^1_\nu/\partial u^{(1)}_o))(t,x(t),\mathbf{U}(t)) \in \mathcal{L}(t)$$

for all $\tau \geq 0$ and $\nu < \mu$, $t \in I$.

That all vectors of the form (18.9),(i) and (ii) belong to $\mathcal{L}(t)$ is then a consequence of (13.10), the same conclusion for the remaining vectors of type (iii) and (iv) can be drawn from Theorem 17.2.

As an application we complete the proof of Lemma 18.3 and determine the sign of $\zeta(\mu,\tau)$.

<u>Corollary</u>. For any given even $\mu > 0$ there exists a particular system $\dot{x} = f(x;u)$ having the following properties (t is again omitted). (i) u is scalar (hence $B_\nu = B_\nu^1$). (ii) For a certain x,μ the linear space $\mathcal{L}$ is spanned by $B_0,\ldots,B_d$, $d = \mu/2$. (iii) (8.19) and all conditions (18.9) are satisfied. (iv) $-(-1)^{\mu/2}\partial B_\mu/\partial u_0 \in P_U(x,\mu)$

The corollary, combined with the weak version (18.11) of Lemma 18.3 yields the strong version. Indeed, let us consider an arbitrary system with satisfies all hypothesis of Lemma 18.3 for some even $\mu \geq 0$. We can then find one and the same real number ζ such that (18.11) holds true both for the given system and for the special system mentioned in the corollary. Because of (iv) it is then clear that the sign of ζ is equal to $(-1)^{\mu/2}$.

<u>Proof of the corollary.</u> Since μ is positive and even this number can be written in the form $2(n-1)$, where n is an integer ≥ 1. Let us now consider the n-dimensional system which is given componentwise as follows

(19.2) $\dot{x}^1 = \frac{1}{2}(x^2)^2, \dot{x}^2 = x^3, \ldots, \dot{x}^{(n-1)} = x^n,\ \dot{x}^n = u$,

u being scalar. It is then easy to see that the first n of the vectors $B_i = B_i(x,\mu)$ are given as

(19.3) $B_i = (-1)^i e_{n-i}$ for $i=0,\ldots,n-2$, $B_{n-1} = (-1)^{n-1}x^2 e_1$,

where e_j is the j-th canonical basis vector (1 at place j, 0 otherwise). Furthermore, from Theorem 17.2 and (19.3) we obtain the relations

$$(-1)^\sigma \sum_{\tau=0}^{\sigma+1} \binom{\sigma+1}{\tau}(-1)^\tau \Gamma^\tau(\partial B_{\sigma+\rho+1-\tau}/\partial u_0) = 0 \quad \text{if} \quad \rho,\sigma < n-1$$

which imply

(19.4) $\partial B_\nu/\partial u_0 = 0$ if $\nu \leq 2n-3$, i.e. if $\nu < \mu$.

Applying the theorem once more we obtain

$$-e_1 = [B_{n-2}, B_{n-1}] = (-1)^{n-2} \partial B_{2(n-1)}/\partial u_o ,$$

in other words

(19.5) $\partial B_\mu/\partial u_o = (-1)^{\mu/2} e_1$.

As one can see immediately from (19.2), $x\equiv 0$, $u\equiv 0$ is a solution of (19.2). We use this solution as our reference solution on some interval $[0,t_1]$ and take as control region U the whole u-space. It follows then from Definition 9.2 (Remark (iii), note the time-independence!)

$\mathcal{P}_U(0,\mathbf{o}) \subseteq \Pi_t$ for all $t \in [0,t_1]$.

Now the reference solution minimizes $x^1(t_1)$ in the class of all solutions satisfying the initial condition $x(0) = 0$.One concludes from the HNC (Theorem 9.1) that there exists a non-vanishing adjoint state variable $y(\cdot)$ such that $y(t)^T p \le 0$ for all $p \in \Pi_t$ and $-y(t_1)^T e_1 \ge 0$. We are thus arrived at the following result: There exists a n-dimensional vector y with these properties

(19.6) $y \neq 0$, $y^T e_1 \le 0$, $y^T p \le 0$ for all $p \in \mathcal{P}_U(0,\mathbf{o})$.

Since the linear space spanned by the B_i is contained in $\mathcal{P}_U(0,\mathbf{o})$ (cf. Corollary 1 to Lemma 14.2) y has to be orthogonal to e_j, $j=2,\dots,n$, in view of (19.3). This statement has to be compatible with the first two inequalities (19.6) and therefore it is obvious that we can assume without loss of generality that $y = -e_1$. It follows then from the third of the inequalities (19.6) and from (19.5) that

(19.7) $-e_1 = -(-1)^{\mu/2} \partial B_\mu/\partial u_o \notin \mathcal{P}_U(0,\mathbf{o})$.

Furthermore it is clear that B_i, $i=0,\dots,n-2$ form a basis of $\mathcal{L}(t)$ for every t, since these vectors represent $n-1 = \mu/2$ linearly independent elements of $\mathcal{L}(t)$ (note that $\mathcal{L}(t)$ is orthogonal to $y=-e_1$ and hence cannot be the whole $\mathbb{R}^n$) It follows therefore from (19.4). and from Lemma 19.1 that - for the given integer μ - the system (19.2) is one which satisfies all conditions (18.9) and which in addition has also the property (iv) stated in the corollary (in view of

(19.7)). Thereby the proof of the corollary is completed.

We add a further lemma of rather elementary nature which we will use frequently in the sequel.

Lemma 19.2. Let $b_\nu(t)$, $\nu=1,2,\ldots,$ be a sequence of n-dimensional vectors which depend continuously upon t for all t in some interval I. Let $\mathcal{L}(t)$ be the linear space spanned by the $b_\nu(t)$ and let $d(t) = \dim\mathcal{L}(t)$.

Claim: There exists a dense subset $\mathcal{S} \subseteq I$ such that $d(\cdot)$ is constant on a neighborhood of each $t \in \mathcal{S}$.

Proof. We define recursively the following sets of real numbers

$$\mathcal{S}_n = \{t \in I : d(t) = n\},$$

$$\mathcal{S}_i = \{t \in I : d(t) = i \text{ and } t \notin \text{closure}\,(\bigcup_{j>i} \mathcal{S}_j)\},$$

$i = n-1, n-2, \ldots, 0$. It is clear that $\mathcal{S} = \bigcup_{i=0}^{n} \mathcal{S}_i$ is dense in I.

Furthermore, if we have a $\tilde{t} \in \mathcal{S}$ then $\tilde{t}$ belongs to some $\mathcal{S}_i$ and this implies that one can find a neighborhood $\tilde{I}$ of $\tilde{t}$ such that $d(t) \le i$ for all $t \in \tilde{I}$. On the other hand $d(\tilde{t}) = i$, i.e. there exist i linearly independent vectors among the $b_\nu(t)$. For reasons of continuity these vectors remain independent for all t sufficiently close to $\tilde{t}$, that is $d(t) \ge i$ on a neighborhood of $\tilde{t}$. It is then also clear that one has $d(t) = \text{const} = i$ on a certain neighborhood of $\tilde{t}$.

Corollary. There exists an open dense subset $\mathcal{S}$ of I such that $\mathcal{L}(t)$ has constant dimension on a neighborhood of each $t \in \mathcal{S}$.

The results established in Sec. 18 and 19 actually provide the proof for the generalized Clebsch-Legendre condition. It merely remains the problem of stating this condition in an appropriate form. This will be done in Sec. 20. We then will draw considerable advantage from the fact that in our considerations so far we did not impose any restrictions - except differentiability - upon $f(t,x,u)$.

For later purposes we have to explain what we mean if we say that a control variable is "updated" to a state variable. Given a control system of the form (7.1), let us pick one scalar control variable appearing in f and call it u. Actually f may depend upon further control variables, but these are omitted for shortness. In accordance with previous custom we denote by u_i, $i=0,1,\ldots$, independent scalar variables and by $\mathbf{u}$ the sequence $\{u_o,u_1,\ldots\}$. Let us now consider the system

$$(19.8)\qquad \dot{x} = f(t,x;u_o),\quad \dot{u}_o = u_1,\ \dot{u}_1 = u_2,\ldots,\ \dot{u}_{k-1} = u_k\ .$$

Here $\hat{x} := (x^T,u_o,\ldots,u_{k-1})^T$ is regarded as the state variable. The role of the control variable in (19.8) is played by the scalar u_k and possibly the control variables different from u which may appear in f. The subsequent lemma relates the relevant quantities for the given system (7.1) to those of the updated system (19.8) (the latter ones are indicated by a hat).
The notation is the usual one. $\hat{e}_i$ is the (n+k)-dimensional unit vector with 1 at place i and zero elsewhere. (n+k)-dimensional vectors are partitioned into a n-dimensional and a k-dimensional vector.

<u>Lemma 19.3.</u> (i) If $\hat{g}(t,x,\mathbf{u}) = (g(t,x,\mathbf{u})^T,0)^T$ then $\hat{\Gamma}(\hat{g})=(\Gamma(g)^T,0)^T$.
(ii) We have

$$\hat{B}_\nu = (-1)^\nu\hat{e}_{n+k-\nu}\ \text{ if }\ \nu<k,\quad \hat{B}_\nu = (-1)^k(B_{\nu-k},0)^T\ \text{ if }\ \nu\geq k\ .$$

(iii) Given a solution $u(\cdot)$, $x(\cdot)$ of (7.1) such that $u(\cdot)$ is of class C^∞ in a neighborhood of $\tilde{t}$ and $u(\tilde{t})\in \operatorname{int}U$. Let Π^*_t, $\hat{\Pi}^*_t$ be the sets associated with this solution according to Definition 9.2 (with respect to the system (19.8) the reference solution is $u_k = u^{(k)}(t),\hat{x}(t) = (x(t),u(t),\dot{u}(t)\ldots))$. Claim: $p = (p^T,*)^T\in \hat{\Pi}^*_{\tilde{t}}$ implies $p\in \Pi^*_{\tilde{t}}$.

<u>Proof.</u> (i) and (ii) are verified immediately in view the definitions given in Sec. 13. Note that the Jacobian matrix of the right side of eq. (19.8) is given as

$$\left(\begin{array}{c|c|ccc} f_x & B_o & & 0 & \\ \hline & & 1 & & \\ 0 & 0 & & \ddots & 1 \\ & & & & 0 \end{array}\right)$$

(iii) follows by inspection from Definition 9.2 (take the initial time t_c sufficiently close to $\tilde{t}$).

20. The generalized Clebsch-Legendre condition.

Let there be given a (not necessarily optimal) reference pair $u(\cdot)$, $x(\cdot)$ on some closed interval I and let $v(\cdot) = (v^1(\cdot),\dots,v^m(\cdot))^T$ be a m-tupel of functions which are of class C^∞ on I. We introduce a new scalar variable ξ and put

$$\tilde{f}(t,x,\xi) = \tilde{f}_{v(\cdot)}(t,x,\xi) \; := f(t,x;u(t)+\xi v(t)).$$

We assume throughout this chapter that the reference control is of class C^∞ on I. Therefore $\tilde{f}$ will become a C^∞function of (t,x,ξ) on a neighborhood $\mathcal{N}$ of the curve $t, x = x(t), \xi=0$, $t\in I$. We now associate with $u(\cdot),x(\cdot),v(\cdot)$ a sequence of n-dimensional vectors $\tilde{b}_\nu = \tilde{b}_\nu(t,x,\xi)$ which are C^∞-functions of (t,x,ξ) and are defined on $\mathcal{N}$. In explicit terms the definition is analogous to the one used for the B_ν (cf. (13.9)) and runs as follows

$$\tilde{b}_o(t,x,\xi) = \frac{\partial}{\partial\xi} f(t,x;u(t)+\xi v(t)),$$

$$\frac{d^\nu}{dt^\nu}(y^T\tilde{b}_o(t,x,\xi)) = y^T\tilde{b}_\nu(t,x,\xi), \quad \nu=1,2,\dots, \tag{20.1}$$

where d/dt means differentiation with respect to the differential equation

$$\dot{x} = f(t,x;u(t)+\xi v(t)), \quad \dot{y} = -f_x^T(t,x;u(t)+\xi v(t))y, \quad \dot{\xi}=0 ,$$

(i.e. ξ has to be regarded as a constant parameter).

As before we denote by $\mathbf{U}(t)$ the sequence (7.5).

<u>Theorem 20.1.</u> Let $u(\cdot),x(\cdot),v(\cdot)$ be as explained above and let $\mathcal{L}(t)$ be the linear space spanned by the columns B^i_ν of the matrices $B_\nu(t,x(t),\mathbf{U}(t))$, $i=1,\dots,m$, $\nu=0,1,\dots$ Assume that the following relations hold : $u(t)\in \mathrm{int}(U)$ for all $t\in I$ and

$$\frac{\partial\tilde{b}_\tau}{\partial\xi}(t,x(t),0) \in \mathcal{L}(t) \tag{20.2}$$

for all t in a dense open subset $\mathcal{S}$ of I and $\tau=0,1,\dots,\mu-1$. Furthermore assume that the number μ is maximal in the sense that there is no open dense subset $\mathcal{S}'\subset I$ such that (20.2) is also true for $\nu = \mu$ and all $t\in\mathcal{S}'$.

Claim: μ is even and

$$(-1)^{\mu/2}\frac{\partial\tilde{b}_\mu}{\partial\xi}(t,x(t),0) \in \Pi_t . \tag{20.3}$$

for all t in a dense open subset of I .

Proof. We begin with three remarks. First one observes that it is sufficient to establish the validity of the relation

$$(20.3')\qquad (-1)^{\mu/2}\,\frac{\partial\tilde{b}_\mu}{\partial\xi}(t,x(t),0)\in P_U(t,x(t),\mathbf{U}(t))$$

for all t in a dense open subset of I. Since $(\partial\tilde{b}_\mu/\partial\xi)(t,x(t),0)$ is a continuous function of t (on I) it is clear that the full statement of the theorem follows then from (20.3') and from previous results (cf. the remarks made in connection with Definition 9. 2.

Secondly we may assume for the purpose of the proof that the identities

$$(20.4)\qquad \begin{aligned} \tilde{b}_\tau(t,x,0) &= B^1_\tau(t,x,\mathbf{U}(t)),\\ (\partial\tilde{b}_\tau/\partial\xi)(t,x,0) &= (\partial B^1_\tau/\partial u_o^{(1)})(t,x,\mathbf{U}(t)) \end{aligned}$$

hold for all τ and all t,x. Indeed let us introduce a new $(m+1)$-dimensional control variable $\hat{v} = (\xi,\underline{v}^T)^T = (\xi,v^1,\ldots,v^m)^T$ via the transformation

$$u\to u(t) + \xi v(t) + \underline{v} = \hat{u}(t,\hat{v}).$$

As a result of this transformation the ref. pair is changed to $\hat{v}=0$, $x=x(\cdot)$ whereas the linear space $\mathcal{L}(t)$ remains unchanged, in view of Corollary 1 to Theorem 15.1. Furthermore, if one calculates for the transformed system the columns B^1_ν (with ξ playing the role of $u_o^{(1)}$) along the reference control $\hat{v}=0$ one arrives at the relation (20.4), simply by comparison of the recursive relations (13.9) and (20.1). On the other hand, if the statement (20.3') has been established for the transformed system it certainly holds also for the original system in view of Corollary 3 to Theorem 15.2. - In passing we note that the vectors $\tilde{b}_\tau(t,x(t),0)$ belong to $\mathcal{L}(t)$, for all τ . This follows from the first one of the relations (20.4) and from what was said above about the invariance of $\mathcal{L}(t)$.

Thirdly it means no loss of generality to assume that the set $\mathcal{S}$ has this property

(20.5) The dimension of $\mathcal{L}(t)$ is constant on a neighborhood of any $t\in\mathcal{S}$.

One simply has to replace - if necessary - $\mathcal{S}$ by its intersection with the set described in Lemma 19.2.

From (20.2), (20.4), and (20.5) the statement of the theorem - with (20.3) replaced by (20.3') - is easily inferred by combining Lemma 18.3 and Lemma 19.1. That μ is an even number can be seen by the following argument. Let us assume that μ is odd. μ is then necessarily positive and this implies, according to Lemma 19.1 and in view of (20.4), that all vectors of the form (18.9), (iv), are elements of $\mathcal{L}(t)$, if $t \in \mathcal{S}$. Taking $\sigma=\rho=(\mu-1)/2$ we see that $\partial B_\mu^1/\partial u_o^{(1)} = \partial \tilde{b}_\mu/\partial \xi$, evaluated along the reference pair, belongs to $\mathcal{L}(t)$ for all $t \in \mathcal{S}$, in contrast to our hypothesis concerning μ.

Corollary 1 (Generalized Clebsch-Legendre condition. First version). Let the reference pair be optimal in the sense explained in Sec. 9 and let the hypotheses of Theorem 20.1 be satisfied. Then there exists an adjoint state vector $y(\cdot)$ which has all properties stated in Theorem 9.1 and hence in particular satisfies the condition

(20.6) $$(-1)^{\mu/2} y(t)^T \frac{\partial \tilde{b}_\mu}{\partial \xi}(t,x(t),0) \le 0$$

for all $t \in I$ (= a subinterval of $[t_o, t_e]$).

There is a different way of expressing both the hypothesis and the conclusion of Theorem 20.1 and its corollary. It is based on the identity

(20.7) $$[B_{\nu-1}^1, B_\nu^1] + (-1)^\nu \partial B_{2\nu}^1/\partial u_o^{(1)} = -(-1)^\nu \sum_{\tau=1}^{\nu} \binom{\nu}{\tau} (-1)^\tau \Gamma^\tau (\partial B_{2\nu-\tau}^1/\partial u_o^{(1)}).$$

(cf. Theorem 17.2. Take $\mu=\nu-1, \rho=\sigma=1$).

Corollary 2. Let $\mathcal{L}(t), \tilde{b}_\nu(t,x,\xi), \nu=0,1,\ldots,$ have the same meaning as in Theorem 20.1. Assume that the following relations hold: $u(t) \in \mathrm{int}(U)$ for all $t \in I$ and

(20.2') $$\frac{\partial \tilde{b}_o}{\partial \xi}(t,x(t),0) \in \mathcal{L}(t), \qquad [\tilde{b}_{\nu-1}, \tilde{b}_\nu](t,x(t),0) \in \mathcal{L}(t)$$

for all t in a dense open subset $\mathcal{S}$ of I and $\nu=1,\ldots,\rho$, where $\rho \ge 0$ (if $\rho=0$ then (20.2') reduces to the first condition). Claim:

(20.3'') $$-[\tilde{b}_\rho, \tilde{b}_{\rho+1}](t,x(t),0) \in \Pi_t + \mathcal{L}(t)$$

for all t in a dense open subset of I. Hence if the reference pair is optimal the inequality

(20.6') $$y(t)^T([\tilde{b}_\rho, \tilde{b}_{\rho+1}](t,x(t),0)) \ge 0$$

holds for all $t \in I$, $y(\cdot)$ being an adjoint state vector which has the properties stated in Theorem 9.1.

For $\rho = 0$ (20.6') is known as Kelley's condition (cf. [5], [7] p.67).

<u>Proof.</u> As before we may assume without loss of generality that (20.4) and (20.5) hold true. Under this assumption the following statements are easily inferred from (20.7), the first assertion of Theorem 20.1 (" μ is even") and the first corollary of Lemma 13.1.

i) $(\partial\tilde{b}_0/\partial\xi)(t,x(t),0) \in \mathcal{L}(t)$ for all t in a dense open subset $\mathcal{S}$ of I implies $(\partial\tilde{b}_1/\partial\xi)(t,x(t),0) \in \mathcal{L}(t)$ for all $t \in \mathcal{S}$.

ii) If (20.2) holds for $\tau=0,\ldots,2\nu-1$, $\nu \geq 1$, and all $t \in \mathcal{S}$ then

(20.8) $$[\tilde{b}_{\nu-1},\tilde{b}_\nu](t,x(t),0) + (-1)^\nu \frac{\partial\tilde{b}_{2\nu}}{\partial\xi}(t,x(t),0) \in \mathcal{L}(t)$$ for all $t \in \mathcal{S}$.

iii) If (20.2) holds for $\tau=0,\ldots,2\nu-1$ and if in addition $[\tilde{b}_{\nu-1},\tilde{b}_\nu](t,x(t),0) \in \mathcal{L}(t)$ for all $t \in \mathcal{S}$ then
$$\frac{\partial\tilde{b}_\tau}{\partial\xi}(t,x(t),0) \in \mathcal{L}(t) \quad \text{for } \tau = 2\nu, 2\nu+1 .$$

It follows then from (i)-(iii) and (20.2') that the condition (20.2) is satisfied for $\tau=0,\ldots,2\rho+1$. This implies - in view of Theorem 20.1 - that (20.3) holds true for $\mu= 2(\rho+1)$ and the desired result (20.3'') then can be inferred immediately from (20.8) for $\nu = \rho+1$.

Though the two versions of the generalized Clebsch-Legendre condition are equivalent the second one offers two advantages if it comes to applications: (i) In order to arrive at the same conclusion one has to take all $\tilde{b}_\nu$ for $\nu \leq 2\rho$ into account if one uses Theorem 20.1, whereas Theorem 20.2 requires the calculation of $\tilde{b}_\nu$ for $\nu \leq \rho+1$ only. The reason for this discrepancy seems to rest upon the fact that the hypothesis of Theorem 20.1 contains redundant information since the validity of condition (20.2) for $\tau=0,\ldots,2\rho$ always implies its validity for $\tau=2\rho+1$. (ii) The computation of $\tilde{b}_\nu$ and $[\tilde{b}_{\nu-1},\tilde{b}_\nu]$ involves differentiation with respect to x but not differentiation with respect to ξ.

The second remark suggests a more convenient formulation of the generalized Clebsch-Legendre condition which avoids the usage of a control parameter ξ . To this purpose let us introduce, for a given control function $u(\cdot)$ and a given m-tupel $v(\cdot) = (v^1(\cdot),\ldots,v^m(\cdot))^T$ of C^∞-functions, a sequence $b_\nu(t,x)$ as follows

$$b_o(t,x) := B_o(t,x,u(t))^T v(t) = f_u(t,x;u(t))^T v(t) ,$$

$$(20.9)\quad b_{\nu+1}(t,x) := \frac{\partial b_\nu}{\partial t}(t,x) + (b_\nu)_x(t,x) f(t,x;u(t)) - $$

$$- f_x(t,x;u(t)) b_\nu(t,x), \quad \nu=0,1,\ldots$$

Theorem 20.2 (Generalized Clebsch-Legendre condition. Second version). Given a reference pair $u(\cdot), x(\cdot)$ and a m-tupel $v(\cdot)=(v^1(\cdot),\ldots,v^m(\cdot))^T$ of C^∞-functions of t. Let the sequence $b_\nu(t,x)$ be defined according to (20.9) and let $\mathcal{L}(t)$ be the linear space defined in Theorem 20.1. Assume that the following conditions hold:
$u(t)\in \mathrm{int}(U)$ for all $t\in I$ (= subinterval of $[t_o,t_e]$)

$$(20.10)\quad \sum_{i,j=1}^{m} f_{u^i u^j}(t,x(t);u(t)) v^i(t) v^j(t) \in \mathcal{L}(t),$$

$$(20.11)\quad [b_{\nu-1},b_\nu](t,x(t)) \in \mathcal{L}(t), \quad \nu=1,\ldots,\rho$$

for all t in a dense open subset $\mathcal{S}$ of I. Here ρ is a non-negative integer (in case $\rho=0$ (20.11) is meaningless). Claim:

$$(20.12)\quad -[b_\rho,b_{\rho+1}](t,x(t)) \in \Pi_t + \mathcal{L}(t)$$

for all t in a dense open subset of I. Hence if the reference pair is optimal the inequality

$$(20.13)\quad y(t)^T([b_\rho,b_{\rho+1}](t,x(t)) \geq 0$$

holds for all $t\in I$, $y(\cdot)$ being an adjoint state vector which has the properties stated in Theorem 9.1.
Proof. Except for the notation the theorem is identical with Corollary 2 to Theorem 20.1.

Next we wish to demonstrate by means of an example that Theorem 20.2 is more suited for applications than the standard form of the generalized Clebsch-Legendre condition. The example is taken from [10] and commonly known as Lawden's spiral.

In accordance with the notation used in [10] we denote the last component of the 5-dimensional state variable by m and the components of the 2-dimensional control variable by T,Θ . T is subject to a restriction $0\leq T\leq T_{Max}$ whereas Θ is free. Hence the condition $u(t)\in \mathrm{int}(U)$ means $0<T(t)<T_{Max}$, and it is this condition which characterizes the singular arcs considered in [10]. We wish to indicate a short and transparent access to the necessary condition which has to hold along a singular arc and which is stated in [10] on p.90.

It is convenient for this purpose to partition the state variable in the form $\hat{x} = (x,z,m)$, where x and z are 2-dimensional and m is scalar. The system equations can then be written in the form

$$(20.14) \quad \dot{x} = z, \; \dot{z} = k(x) + (T/m)Q(\theta), \; \dot{m} = -T/c$$

(cf. [10], p. 92. Our x,z respectively correspond to the variables which are denoted there by (y,x) and (u,v) respectively). The 2-dimensional vector k depends upon the state variable x only, the 2-dimensional vector Q upon the control variable θ only. The explicit form of k,Q does not matter for our purpose.
Since the equations are linear in the control variable T the expression on the left hand side of (20.10) will become identically zero if we choose $v(t) = (1,0)^T$. Hence (20.10) is then trivially satisfied.
Let now there be given a fixed reference control $u(t) = (T(t),\theta(t)) \in \mathrm{int}(U)$. We take $v(t) = (1,0)^T$ and have then

$$b_o(t,\hat{x}) = (0, \tfrac{1}{m} q(t), -\tfrac{1}{c}) \text{ where } q(t) := Q(\theta(t)).$$

Next we calculate b_1, b_2 via the recursive relations (20.9). Note that in the present case the role of x is played by the triple (x,z,m) and the role of $f(t,x;u(t))$ by the right hand side of (20.14) taken along the reference control, i.e. $f(t,x;u(t))$ is the right hand side of the system

$$(20.14') \quad \dot{x} = z, \quad \dot{z} = k(x) + \frac{1}{m}\tilde{q}(t), \; \dot{m} = \zeta(t)$$

where

$$(20.15) \quad \tilde{q}(t) := T(t)q(t) \quad \text{and} \quad \zeta(t) := -T(t)/c.$$

One finds then

$$b_1 = \begin{pmatrix} 0 \\ \frac{1}{m}\dot{q} \\ 0 \end{pmatrix} + \begin{pmatrix} 0 & 0 & 0 \\ 0 & 0 & -\frac{1}{m^2} q \\ 0 & 0 & 0 \end{pmatrix} \begin{pmatrix} z \\ k + \frac{1}{m}\tilde{q} \\ \zeta \end{pmatrix}$$

$$- \begin{pmatrix} 0 & E & 0 \\ k_x & 0 & -\frac{1}{m^2}\tilde{q} \\ 0 & 0 & 0 \end{pmatrix} \begin{pmatrix} 0 \\ \frac{1}{m} q \\ -\frac{1}{c} \end{pmatrix}$$

Hence

$$(20.16) \quad b_1(t,\hat{x}) = \frac{1}{m} \begin{pmatrix} -q(t) \\ \dot{q}(t) \\ 0 \end{pmatrix} .$$

(Note that $q(t)\zeta(t) + \frac{1}{c}\tilde{q}(t) = 0$ because of (20.15)).

$$b_2 = \frac{1}{m}\begin{pmatrix} -\dot{q} \\ \ddot{q} \\ 0 \end{pmatrix} + \begin{pmatrix} 0 & 0 & \frac{1}{m^2} q \\ 0 & 0 & -\frac{1}{m^2}\dot{q} \\ 0 & 0 & 0 \end{pmatrix} \cdot \begin{pmatrix} z \\ k+\frac{1}{m}\tilde{q} \\ \zeta \end{pmatrix}$$

$$- \frac{1}{m}\begin{pmatrix} 0 & E & 0 \\ k_x & 0 & -\frac{1}{m^2}\tilde{q} \\ 0 & 0 & 0 \end{pmatrix} \cdot \begin{pmatrix} -q \\ \dot{q} \\ 0 \end{pmatrix}$$

(20.17) $$b_2 = \frac{1}{m}\begin{pmatrix} -\dot{q} \\ \ddot{q} \\ 0 \end{pmatrix} + \zeta\frac{1}{m^2}\begin{pmatrix} q \\ -\dot{q} \\ 0 \end{pmatrix} - \frac{1}{m}\begin{pmatrix} \dot{q} \\ -k_x q \\ 0 \end{pmatrix} .$$

Hence

$$b_2(t,\hat{x}) = b_2'(t,\hat{x}) + \frac{1}{m}\begin{pmatrix} -\dot{q}(t) \\ k_x(x)q(t) \\ 0 \end{pmatrix} ,$$

where b_2 is the sum of the first two vectors which appear on the right hand side of (20.17). It is immediately verified that b_1 and b_2' commute: $[b_1, b_2'] = 0$. Hence we have, from (20.16), (20.17),

(20.18) $$[b_1,b_2](t,\hat{x}) = -\frac{1}{m}\begin{pmatrix} 0 \\ \left(k_x(x)q(t)\right)_x q(t) \\ 0 \end{pmatrix} .$$

One also finds by straightforward calculation from (20.16) and from the definition of b_o that

$$[b_o,b_1] = \frac{1}{cm} b_1 .$$

This implies - in view of one remark preceding the proof of Theorem 20.1 - that $[b_o,b_1] \in \mathcal{L}(t)$ along any reference pair and hence the hypotheses (20.10), (20.11) of Theorem 20.2 are satisfied with $\rho=1$. The conclusion of the Theorem is then that the inequality

(20.19) $$\hat{y}(t)^T([b_1,b_2](t,\hat{x}(t))) \geq 0$$

has to hold along any possible singular arc, where $\hat{y}(\cdot)$ is an adjoint vector for the optimization problem in question. If $\hat{y}$ is partitioned in the same way as $\hat{x}$ and if the second component is denoted by y then (20.18) and (20.19) lead to the inequality

(20.20) $y(t)^T (k_x(x)q(t))_x q(t) \le 0$ for $x=x(t)$.

In the foregoing analysis we never used the fact that x is 2-dimensional and that θ is 1-dimensional. So actually (20.20) is a necessary condition which is valid along singular extremals for any second order system of the form

$$\ddot{x} = k(x) + (T/m)Q(\theta), \quad \dot{m} = -T/c$$

where m,T are scalars, x and θ arbitrary vectors and where the control variable θ assumes values in an open set.
Finally we note that (20.20) can be written in a more familiar way if one uses the scalar function

$$H^*(x,y) = y^T k(x) \; .$$

Indeed, if x is of dimension r and if $Q^i(\theta)$ are the components of the r-dimensional vector $Q(\theta)$ then the left hand side of (20.20) is nothing else than

$$\sum_{i,j=1}^{r} \frac{\partial^2 H^*}{\partial x^i \partial x^j} (x,y) Q^i(\theta) Q^j(\theta)$$

calculated along the reference pair. In the case of Lawden's spiral this quantity assumes positive values thus ruling out the possibility of a minimizing arc.

21. Multivariable controls. Second order equality conditions.

We will consider in this section control systems depending upon two scalar control variables, i.e. $u=(u^1,u^2)$ is 2-dimensional. One could as well admit systems of arbitrary control dimension $m \ge 2$ and then pass to a system of the form

(21.1) $\dot{x} = f(t,x;u(t) + \xi_1 v_1(t) + \xi_2 v_2(t))$

where ξ_1, ξ_2 are then formally regarded as control variables. This approach would be the obvious analogue to the setting for the Clebsch-Legendre condition employed in the previous section. Indeed all results stated in the following could be phrased in such a way that the actual control dimension m plays no role. The vectors B^1_ν, B^2_ν which appear in the sequel have then to be calculated from the system (21.1) with ξ_1, ξ_2 playing the role of u^1, u^2. However the linear space

$\mathcal{L}(t)$ is the linear space associated with the given m-dimensional system rather than with the 2-dimensional system (21.1) (the latter one would in general give rise to a smaller space). In order to avoid too lengthy formulations however we restrict ourselves to the case $m=2$. $B_\nu = (B_\nu^1, B_\nu^2)$ is then a matrix consisting of two n-dimensional column vectors. Our main tool will again be the formula of Theorem 17.2, which is repeated here for the reader's convenience:

$$(21.2)\quad [B_\mu^i, B_\nu^j] = (-1)^\mu \sum_{\tau=0}^{\mu+1} \binom{\mu+1}{\tau} (-1)^\tau \Gamma^\tau (\partial B_{\nu+\mu+1-\tau}^i / \partial u_o^j) .$$

As before, let $u(\cdot)$, $x(\cdot)$ be a fixed reference pair which need not to be optimal, unless otherwise stated. $\mathcal{U}(t)$ again is the sequence (7.5) and $\mathcal{L}(t)$ the linear space spanned by the columns of the matrices $B_\nu(t,x(t),\mathcal{U}(t))$. I is a subinterval of $[t_o,t_e]$.

Lemma 21.1. Assume that the following quantities, evaluated at $(t,x,\mathcal{U}) = (t,x(t),\mathcal{U}(t))$, belong to $\mathcal{L}(t)$ for all t in an open dense subset $\mathcal{S}$ of I:

$$(21.3)\qquad \partial B_o^1/\partial u_o^2 ,\quad [B_\nu^1, B_\nu^2] \text{ for } 1 \leq \nu < \rho ,\quad [B_{\nu-1}^1, B_\nu^2] \text{ for } 1 \leq \nu \leq \rho .$$

Claim: (i) $[B_\mu^1, B_\sigma^2](t,x(t),\mathcal{U}(t)) \in \mathcal{L}(t)$ if $\mu+\sigma < 2\rho$,

(ii) $(\partial B_\lambda^1/\partial u_o^2)(t,x(t),\mathcal{U}(t)) \in \mathcal{L}(t)$ if $\lambda \leq 2\rho$

on some open dense subset of I.

Proof. As before we may assume without loss of generality that the set $\mathcal{S}$ has the property (20.5). It follows then from the first corollary to Lemma 13.1 that application of the operators $\Gamma^\rho, \rho \geq 0$, to the vectors (21.3) generates again elements of $\mathcal{L}(t)$. Consider now the sequence of relations obtained from (21.2) by specializing the indices as follows: $i=1$, $j=2$,

$(\mu,\nu) = (0,0),(0,1),(1,1),(1,2),(2,2),\ldots,(\rho-1,\rho)$.

The second conclusion of the lemma follows then immediately from the preceding remark. Using (21.2) once more - direction reversed - one arrives at the first conclusion.

As one observes the statement (i) of the lemma is symmetric with respect to the two control variables u^1, u^2, whereas the hypothesis is not. This remark leads to a further result.

Corollary. If all vectors (21.3) and in addition the vector $\partial B_o^2/\partial u_o^1$ are contained in $\mathcal{L}(t)$ for all $t \in \mathcal{S}$, then we have also

$$(\partial B_\lambda^2/\partial u_o^1)(t,x(t),\mathbf{U}(t)) \in \mathcal{L}(t) \quad \text{if} \quad \lambda \leq 2\rho \;.$$

The next two theorems contain the main results of this sections. The arguments used in the proofs are so similar, that we give all details in the first case only.

Theorem 21.1 Let $u(t) \in \mathrm{int}U$ for all t in a subinterval I of $[t_o,t_e]$ and let the following quantities, evaluated along $(t,x\,\mathbf{U}) = (t,x(t),\mathbf{U}(t))$, be contained in $\mathcal{L}(t)$ for all $t \in I$:

$$\partial B_o^1/\partial u_o^1,\; [B_{\nu-1}^1,\; B_\nu^1],\; \nu=1,\ldots,\rho,\; \rho \geq 0 \;.$$

Claim: If the reference pair is optimal then the relation

$$(21.4) \quad y(t)^T(\partial B_\rho^i/\partial u_o^j)(t,x(t),\; \mathbf{U}(t)) = 0,\; i \neq j,$$

holds for all $t \in I$, $y(\cdot)$ being an adjoint state vector which has the properties stated in Theorem 9.1.

Proof. We first remark that it suffices to prove the theorem in case $i=2$, $j=1$, i.e. to establish the equality

$$(21.4') \quad y(t)^T(\partial B_\rho^2/\partial u_o^1)(t,x(t),\mathbf{U}(t)) = 0.$$

In fact, if the hypothesis of the theorem holds for some fixed $\rho=\rho'$ it holds for all $\rho \leq \rho'$. Hence (21.4') is also true for all $\rho \leq \rho'$. Exploiting the symmetry of the relation (21.2) one then infers the correctness of (21.4) for $i=1$, $j=2$ and $\rho \leq \rho'$.

We may also assume without loss of generality that $\mathcal{L}(t)$ has constant dimension on I. Indeed, if the statement in question has been established under this proviso one can refer to arguments used in the proof of Theorem 20.1 and conclude that (21.4') holds true for $t \in \mathcal{S}$ and, by continuity arguments, then also for $t \in I$. If $\mathcal{L}(t)$ has constant dimension then it is clear - in view of (21.2) for $i=j=1$ and the remarks made in connection with (20.8) - that these quantities, evaluated at $t, x=x(t)$, $\mathbf{U} = \mathbf{U}(t)$, belong to $\mathcal{L}(t)$:

$$(21.5) \quad [B_\nu^1,B_\mu^1] \quad \text{for} \quad \nu+\mu \leq 2\rho,\; \partial B_\lambda^1/\partial u_o^1 \quad \text{for} \quad \lambda \leq 2\rho+1 \quad .$$

We now fix $t \in I$ and put $\mathbf{U} = \mathbf{U}(t)$. Let d be as before a positive integer such that $\mathcal{L}(t)$ is spanned by the columns of E_ν for $\nu<d$. Next we fix positive integers κ, K satisfying the conditions

$$(21.6) \quad 2\rho+1 \leq 2\kappa+2-2d \;,\quad \kappa>2\rho,\; K \geq \kappa,\; K \geq 2\rho+1 \;,\; 2\kappa+2\rho+1<3\kappa+1$$

After that we choose a positive integer M, real numbers $z_1,\dots,z_M$ and real numbers $\eta_o = 0,\eta_1,\dots,\eta_M$ such that the system of linear equations in the M unknowns $\xi_1,\dots,\xi_M$ which in explicit terms is given as

$$\mathcal{L}^{(\nu)}(\underline{\xi}) := \sum_{i=1}^{M} (\xi_{i-1}-\xi_i) z_i^{\nu} = 0, \qquad \nu = 0,\dots,K,$$

(21.7)
$$P^{(\nu)}(\underline{\xi},\underline{\eta}) := \sum_{i=1}^{M} (\xi_{i-1}-\xi_i)(\eta_{i-1}-\eta_i) z_i^{\nu} = 0, \qquad \nu = 0,\dots,K,$$

$$Q^{(\nu,\tau)}(\underline{\xi},\underline{\eta}) := \sum_{1\le i<j\le M} (\xi_{i-1}-\xi_i)(\eta_{j-1}-\eta_j) z_i^{\nu} z_j^{\tau} = 0, \qquad \nu,\tau=1,\dots,K,$$

and

(21.8)
$$S^{(\nu)}(\underline{\xi},\underline{\eta}) := \sum_{i=1}^{M} (\xi_{i-1}-\xi_i)(\eta_{i-1}+\eta_i) z_i^{\nu} = 0, \qquad \nu =0,\dots,K,$$

$$S^{(\nu)}(\underline{\eta},\underline{\xi}) := \sum_{i=1}^{M} (\eta_{i-1}-\eta_i)(\xi_{i-1}+\xi_i) z_i^{\nu} = \begin{cases} 0 & \text{if } \nu \ne 2\rho+1, \\ \beta & \text{if } \nu = 2\rho+1, \end{cases}$$

$\nu = 0,\dots,K$, admits for every β a solution $\underline{\xi}=\underline{\xi}(\beta)$. By the symbols $\underline{\xi}(\beta),\underline{\eta}$ respectively we denote the $(M+1)$-tuples

(21.9) $(0,\xi_1(\beta),\dots,\xi_M(\beta))^T$ and $\underline{\eta} = (0,\eta_1,\dots,\eta_M)^T$

respectively, the first components ξ_o,η_o being 0. Actually we have to require more than the independence of the linear forms (in $\underline{\xi}$) which appear on the right hand sides of (21.7), (21.8). M, z_i,η_i should be chosen in such a way that also the conditions (21.10) below are satisfied. That all this can be done follows from some elementary considerations which are explained in the appendix (Lemma 2.5). The additional conditions imposed on $z_i,\underline{\eta}$ are

(21.10)
$$z_1<z_2< \dots <z_M<0,$$
$$\mathcal{L}^{(\nu)}(\underline{\eta}) := \sum_{i=1}^{M} (\eta_{i-1}-\eta_i) z_i^{\nu} = 0 \quad \text{for } \nu = 0,\dots,K.$$

For later purposes we write down a relation which is an immediate consequence of (21.7)-(21.10), namely

(21.11) $$Q^{(\nu,\mu)}(\underline{\xi},\underline{\eta}) + Q^{(\mu,\nu)}(\underline{\eta},\underline{\xi}) + P^{(\nu+\mu)}(\underline{\xi},\underline{\eta}) = \mathcal{L}^{(\nu)}(\underline{\xi})\mathcal{L}^{(\mu)}(\underline{\eta}) = 0$$

if $\mu \le K$.

We now introduce 2-dimensional vectors $v_{i,j}(\beta)$, $i=1,\dots,M$, $j=1,2,\dots$, as follows [1)]

$$v_{i,o}(\beta) = \begin{cases} (\xi_i(\beta),\ 0) & \text{if } \rho > 0, \\ (\xi_i(\beta), \eta_i) & \text{if } \rho = 0, \end{cases}$$

(21.12)

$$v_{i,\rho}(\beta) = \begin{cases} (0, \eta_i) & \text{if } \rho > 0, \\ \text{as above} & \text{if } \rho = 0, \end{cases}$$

and $v_{i,j}(\beta) = (0,0)$ for $j \neq 0,\rho$. Note that $\mathcal{L}^{(o)}(\underline{\xi}) = \xi_M - \xi_o = \xi_M$, hence the relation (21.7),(21.10) implies $v_{o,j} = v_{M,j} = 0$. Let then the sequences $\mathcal{V}_i = \mathcal{V}_i(\beta)$ and $\mathcal{U}_i = \mathcal{U}_i(\beta,\lambda)$ be defined as

$$\mathcal{V}_i = \{v_{i,o}, v_{i,1}, \dots\} , \quad \mathcal{U}_i = \mathcal{U} + \lambda^{\kappa} \mathcal{V}_i , \quad i=0,\dots,M.$$

Since $\mathcal{V}_o = \mathcal{V}_M = \mathcal{O}$ we have

(21.13) $$\mathcal{U}_o = \mathcal{U}_M = \mathcal{U}(= \mathcal{U}(t)).$$

For the remaining portion of the proof the quantities $t, x=x(t)$, $\mathcal{U} = \mathcal{U}(t) = \mathcal{U}_o$, $\mathcal{V}_i$ and $z=(z_1,\dots,z_M)^T$ have to be regarded as fixed whereas λ and β are scalar variables. We wish to study the formal power series in λ which are defined in terms of the power series introduced in Sec. 8 and Sec. 14 as follows

(21.14)
$$\tilde{C}(\lambda) := C(t,x(t),\lambda z , \mathcal{U}_o,\dots,\mathcal{U}_M) ,$$
$$\tilde{C}^{(i)}(\lambda) := C^{(i)}(t,x(t),\lambda z, \mathcal{U}_o,\dots, \mathcal{U}_M)$$

$i=1,2$. To this purpose we first compute $l^{(\nu)}$ and $L^{(\nu)}$ for the present choice of $\mathcal{U}_i$ (cf. (21.12)). The formulas below can be derived immediately from Definition 12.1 and Theorem 13.1. The first one is an identity in the variable x, the subsequent ones are needed only along the reference solution, hence the argument $t, x=x(t)$ and in part also $\mathcal{U}_o = \mathcal{U}(t)$ is omitted. We also write $\underline{\xi}$ instead of $\underline{\xi}(\beta)$.

(21.15)
$$l^{(\nu)}(t,x; \mathcal{U}_o, \mathcal{U}_{i-1} - \mathcal{U}_i) = \lambda^{\kappa}\{ (\xi_{i-1}-\xi_i) B^1_{\nu-1}(t,x,\mathcal{U}_o) + $$
$$+ \binom{\nu-1}{\rho} (\eta_{i-1}-\eta_i) B^2_{\nu-1-\rho}(t,x,\mathcal{U}_o)\} .$$

---

1) ρ is the integer appearing in the hypothesis of the theorem

$$L^{(\nu)}(U_o, U_{i-1}-U_i)\cdot(U_{i-1}+U_i-2\,U_o) =$$

$$= \lambda^{2\kappa}\{(\xi_{i-1}^2-\xi_i^2)(\partial B_{\nu-1}^1/\partial u_o^1) + (\xi_{i-1}-\xi_i)(\eta_{i-1}+\eta_i)(\partial B_{\nu-1}^1/\partial u_\rho^2) +$$

$$+ \binom{\nu-1}{\rho}(\eta_{i-1}-\eta_i)(\xi_{i-1}+\xi_i)(\partial B_{\nu-1-\rho}^2/\partial u_o^1) + \binom{\nu-1}{\rho}(\eta_{i-1}^2-\eta_i^2)(\partial B_{\nu-1-\rho}^2/\partial u_\rho^2)\} .$$

Hence

$$\sum_{i=1}^N l^{(\nu)}(U_o, U_{i-1}-U_i)z_i^\nu = \lambda^\kappa\left(\mathcal{L}^{(\nu)}(\underline{\xi})B_{\nu-1}^1 + \binom{\nu-1}{\rho}\mathcal{L}^{(\nu)}(\underline{\eta})B_{\nu-1-\rho}^2\right) ,$$

$$\sum_{i=1}^N L^{(\nu)}(U_o, U_{i-1}-U_i)(U_{i-1}+U_i-2\,U_o)z_i^\nu =$$

$$= \lambda^{2\kappa}\{S^{(\nu)}(\underline{\xi},\underline{\xi})(\partial B_{\nu-1}^1/\partial u_o^1) + S^{(\nu)}(\underline{\xi},\underline{\eta})(\partial B_{\nu-1}^1/\partial u_\rho^2) +$$

$$+ \binom{\nu-1}{\rho}S^{(\nu)}(\underline{\eta},\underline{\xi})(\partial B_{\nu-1-\rho}^2/\partial u_o^1) + \binom{\nu-1}{\rho}S^{(\nu)}(\underline{\eta},\underline{\eta})(\partial B_{\nu-1-\rho}^2/\partial u_\rho^2)\}.$$

If $\nu \leq K$ the expressions on the right hand side can be simplified in view of (21.7), (21.8) and (21.11). Also certain terms - which may depend upon β - can be identified with elements of the linear space $\mathcal{L}(t)$. In the sequel we will use always the same symbol b or b_ν in order to denote those elements. With z replaced by λz (cf.(21.14)) one verifies immediately from (14.3) and from the fact that $\mathcal{L}^{(\nu)}(\underline{\xi}) = \mathcal{L}^{(\nu)}(\underline{\eta}) = 0$ for $\nu \leq K$ the validity of the following relations

$$\sum_{i=1}^M l^{(\sigma)}(U_o, U_{i-1}-U_i)z_i^\sigma = 0 \quad \text{if} \quad \sigma \leq K,$$

(21.16) $$\tilde{c}^{(1)}(\lambda) = x + \lambda^{2\kappa+1}\sum_{\nu=0}^\infty \lambda^\nu b_\nu .$$

(Note that $K \geq \kappa$, according to (21.6)). Taking into account what has been said about the Lie-brackets $[B_\nu^1, B_\mu^1]$ (cf. (21.5)) one infers from (21.15), from the third of the equations (21.7) and from (14.6) that

(21.17) $$R_{\sigma,\mu} = \lambda^{2\kappa+\sigma+\mu}b \quad \text{if} \quad \sigma+\mu \leq 2\rho+1.$$

Note that $\sigma+\mu \leq 2\rho+1$ implies $\sigma+\mu \leq K$, because of (21.6). Hence the coefficients of all Lie-brackets $[B^1_{\cdot\cdot}, B^2_{\cdot\cdot}]$ vanish in view of (21.7) and (21.11), since they are either of the form $Q^{(\sigma,\mu)}(\underline{\xi},\underline{\eta})$ or of the form $Q^{(\sigma,\mu)}(\underline{\eta},\underline{\xi})$. It remains the problem of evaluating the part of $C^{(2)}$ which is represented by the infinite series appearing on the right hand side of (14.3). Here we have to exploit the relations (21.8) and our hypothesis concerning the derivatives $\partial B^1_\lambda/\partial u^1_o$ (cf.(21.5)). Note also that $\partial B^2_\lambda/\partial u^2_\rho = 0$ if $\lambda<\rho$ (cf. (13.8)). It is then not difficult to see from the above formula for $\sum_i L^{(\nu)})\ldots)\cdot(\ldots)$ that the series in question assumes the form

$$\lambda^{2\kappa}\sum_{\nu=1}^{2\rho}\lambda^\nu b_\nu + \lambda^{2\kappa+2\rho+1}(b+\beta g+h+\mathcal{O}(\lambda))$$

where

$$(21.18) \quad g = \frac{1}{2}\frac{1}{(2\rho+1)!}\binom{2\rho}{\rho}(\partial B^2_\rho/\partial u^1_o)$$

and

$$(21.18') \quad h = \frac{1}{2}\frac{1}{(2\rho+1)!}\binom{2\rho}{\rho}P^{(2\rho+1)}(\underline{\eta},\underline{\eta})(\partial B^2_\rho/\partial u^2_\rho).$$

Combining the last result with (21.16), (21.17) and taking (14.3) and the relation

$$\tilde{C}(\lambda) = \tilde{C}^{(2)}(\lambda) + \mathcal{O}(\lambda^{3\kappa+1})$$

into account one finally arrives at an asymptotic formula of the type (14.10) with

$$r = 2\kappa+1, \quad s = 2\kappa+2\rho+1, \quad p = \beta g+h.$$

Note that - because of $s<3\kappa+1$ (cf.(21. 6)) - the relevant terms of $\tilde{C}(\lambda)$ are provided by $\tilde{C}^{(2)}(\lambda)$. We now are in a position to apply the second corollary to Lemma 14.2. The hypothesis (14.7) is satisfied because of the first inequality (21.6); (14.9) is a consequence of (21.13). The conclusion of the corollary in the present situation then states that

$$(21.19) \quad \beta g+h \in \mathcal{P}_U(t,x(t),\mathcal{U}(t)) ,$$

and it holds true for any fixed but arbitrary $t\in I$ and any β. g and h however are piecewise continuous functions of t and are indepen - dent of β, as can be seen immediately from (21.18) and (21.18'). The desired result (21.4') follows now from Definition 9.2 (Remark (iii)) and Theorem 9.1: (21.19) namely implies $\beta g+h\in \mathcal{T}_t$ and this in turn leads to the inequality $\beta y^T g+y^T h\leq 0$. Since β is arbitrary we necessarily have $y^T g=0$ and this statement finally is nothing else than (21.4').

We wish to add some remarks which might be useful for applications. First we observe that in case $\rho=0$ the statement of the theorem can be expressed in terms of the Hamiltonian as follows (and thereby turns out to be a simple consequence of the Clebsch-Legendre condition): If $\partial^2 H/(\partial u^1)^2$ vanishes along the reference solution, then so does $\partial^2 H/\partial u^1 \partial u^2$.

In case $\rho>0$ there is an alternative version of the theorem which is analogous to the one given for the generalized Clebsch-Legendre condition. It can be derived from (21.2) and the above remark about the case $\rho=0$ by a straightforward application of the definition (13.2) of the operator Γ.

Corollary 1. Let the hypotheses of Theorem 21.1 be satisfied. Then there exists an adjoint state vector which satisfies the conditions of Theorem 9.1 and is orthogonal (along the reference solution) to the Lie-brackets

$$[B^1_\nu, B^2_\mu] \quad \text{for} \quad \nu+\mu \leq \rho-1.$$

Finally we wish to repeat our result for $\rho=1$, since this case seems to be of particular interest for applications.

Corollary 2. If the reference solution is optimal and if the quantities $\partial B^1_o/\partial u^1$, $[B^1_o, B^1_1]$ (evaluated along the solution) are contained in $\mathcal{L}(t)$ for all $t \in I$ then we have an equality type necessary condition which can be written in the following two equivalent forms

$$\text{(i)} \quad \frac{\partial}{\partial u^i} \frac{d}{dt} \frac{\partial H}{\partial u^j} = 0, i \neq j, \qquad \text{(ii)} \quad y^T [B^1_o, B^2_o] = 0 .$$

A particular feature of Theorem 21.1 lies in the fact that nothing is assumed about the derivative $\partial B^2_o/\partial u^2$. If however further conditions are added to the hypothesis such that it becomes symmetric with respect to the components of the control variable one is able to establish a larger set of equality type conditions (cf. [3]).

Theorem 21.2 Let the following quantities, evaluated along the reference solution, be contained in $\mathcal{L}(t)$, for all $t \in I$ and $i=1,2$

$$\partial B^i_o/\partial u^i, \; [B^i_{\nu-1}, B^i_\nu] \; , \; \nu=1,\ldots,\rho_i \geq 0 .$$

Again I is a subinterval of $[t_o, t_e]$ such that $u(t) \in \text{int} U$ for all $t \in I$.

Claim: If the reference solution is optimal then there exists an adjoint state variable $y(\cdot)$ which has the properties described in Theorem 9.1 and is orthogonal to the vectors

(i) $\partial B_\nu^i/\partial u_o^j$ if $i \neq j$ and $\nu \leq \rho_1+\rho_2+1$,

(ii) $[B_\nu^1, B_\mu^2]$ if $\nu+\mu \leq \rho_1+\rho_2$.

<u>Proof.</u> Our aim is to establish the relation

$$(21.20) \qquad \pm(\partial B^2_{\rho_1+\rho_2+1}/\partial u_o^1)(t,x(t),\mathbf{U}(t)) \in \Pi_t$$

for all $t \in \mathcal{S}$, where $\mathcal{S}$ is open and dense in I and $\mathcal{L}(t)$ has constant dimension in a neighborhood of any point of $\mathcal{S}$. Once (21.20) has been shown it is clear we also have

$$\pm \partial B_\nu^2/\partial u_o^1 \in \Pi_t \quad \text{for} \quad 0<\nu \leq \rho_1+\rho_2+1 \ .$$

(Note that ν can be written as $\rho_1'+\rho_2'+1$ with $\rho_i' \leq \rho_i$ and that the hypothesis of the theorem remains of course true if ρ_i is replaced by $\rho_i' \leq \rho_i$). That y is orthogonal to the vectors of the type (i) and for $i=2, j=1$ follows then from Theorem 9.1 (in case $\nu=0$ it is a consequence of the Clebsch-Legendre condition as we have remarked before). In order to complete the proof of the theorem one simply has to make use of the identity (21.2).
Let us assume that $\rho_1 \geq \rho_2$. We put $\rho=\rho_1-\rho_2$ and denote as before by d a positive number which is so large that the space $\mathcal{L}(t)$ is spanned by B_ν for $\nu \leq d$ ($t \in \mathcal{S}$ from now on is regarded as fixed). Next we choose positive integers κ, K such that

$$(21.6') \qquad 2\rho_1+2 \leq 2\kappa+2-d,\ \kappa>2\rho_1+1,\ K \geq \kappa, K \geq 2\rho_1+2\ ,\ 2\kappa+2\rho_1+2<3\kappa+1$$

The essential step in the proof is the determination of N-tuples $\underline{\eta}, \underline{\xi}(\beta)$; this is done in the same way as in the proof of Theorem 21.1 except for the linear equations (21.8) which now assume the form

$$(21.8') \qquad \begin{aligned} &S^{(\nu)}(\underline{\xi},\underline{\eta}) = 0 \qquad , \ \nu=0,\ldots,K, \\ &S^{(\nu)}(\underline{\eta},\underline{\xi}) = \begin{cases} 0 & \text{if } \nu \neq 2\rho_1+2 \ , \\ \beta & \text{if } \nu = 2\rho_1+2. \end{cases} \end{aligned}$$

Having defined $\underline{\xi},\underline{\eta}$ we introduce $\mathcal{U}_i = \mathcal{U}_i(\beta,\lambda)$ by precisely the same formula as we did in the proof of Theorem 21.1. We proceed then also as before and exploit the hypothesis of the theorem. Now we have instead of (21.5) the following relations at our disposal (i=1,2)

$$\begin{aligned} &[B^i_\nu, B^i_\mu] \in \mathcal{L}(t) \quad \text{if} \quad \nu+\mu \leq 2\rho_i\,, \\ (21.5') \quad &\partial B^i_\lambda/\partial u^i_o \in \mathcal{L}(t) \quad \text{if} \quad \lambda \leq 2\rho_i+1\,, \\ &\partial B^2_{\lambda+\rho}/\partial u^2_\rho \in \mathcal{L}(t) \quad \text{if} \quad \lambda \leq 2\rho_2+1\,. \end{aligned}$$

Note that the third line is a consequence of the second one, in view of Lemma 13.2. Combining (21.5') and (21.15) one sees that we have a relation for $R_{\sigma,\mu}$ similar to the previous one:

$$R_{\sigma,\mu} = \lambda^{2\kappa+\sigma+\mu} b, \; b \in \mathcal{L}, \text{ if } \sigma+\mu \leq 2\rho_1+2.$$

So finally, if one uses the formula for $\sum_i L^{(\nu)}(\ldots)\cdot(\ldots)$ as given in the proof of Theorem 21.1 and if one takes (21.8') into account, one sees that in the present situation the formal power series (21.17) assumes the form

$$(21.17') \qquad \lambda^{2\kappa} \sum_{\nu=1}^{2\rho_1+1} \lambda^\nu b_\nu + \lambda^{2\kappa+2\rho_1+2}(b+\beta g+\mathcal{O}(\lambda))\,,$$

where

$$g = \frac{1}{2}\,\frac{1}{(2\rho+2)!}\binom{2\rho+1}{\rho}(\partial B^2_{\rho_1+\rho_2+1}/\partial u^1_o)\,.$$

The remaining portion of the proof follows the familiar pattern. We use (21.6') instead of (21.6) but otherwise the previous arguments and infer from the second corollary of Lemma 14.2 that

$$\beta g \in \Pi_t\,.$$

Since β is an arbitrary real number this is nothing else than the desired result (21.20).

The following corollaries concern inequalities for doubly singular arcs which fit into the context of Sec. 21 because of the hypotheses under which they can be established. It is required that certain vectors which are orthogonal to y according to the last theorems belong to $\mathcal{L}(t)$. In case $\rho=1$ the statement of Corollary 2 has first been observed by Goh [9].

Corollary 1. Let the hypotheses of Theorem 21.2 hold for $\rho_1=\rho_2=\rho>1$. Assume furthermore that these vectors for $t\in I$ belong to $\mathcal{L}(t)$:

$$\partial B_o^i/\partial u_o^j\ ,\ i\neq j,\quad [B^1_{\nu-1},B^2_\nu] + [B^2_{\nu-1},B^1_\nu],\ \nu=1,\dots,\rho.$$

Claim: The quadratic form in the scalar variables ξ,η which is given as

$$y^T[\xi B^1_\rho + \eta B^2_\rho, \xi B^1_{\rho+1} + \eta B^2_{\rho+1}]$$

is positiv-semidefinite on the singular arc (i.e. for $t\in I$).

Proof. Apply Theorem 20.2 to the sequence $\xi B^1_\nu + \eta B^2_\nu$, $\nu = 0,1,\dots$

Corollary 2. Let the hypotheses of Theorem 21.1 hold with $\rho>1$. Assume in addition that these vectors belong to $\mathcal{L}(t)$, for all $t\in I$:

$$\partial B^1_\nu/\partial u^2_o\ ,\quad \nu\leq\rho.$$

Claim: The following quadratic form is positiv-semidefinite for $t\in I$:

$$y^T(\xi^2[B^1_\rho,B^1_{\rho+1}] - 2\xi\eta(-1)^\rho[B^1_\rho,B^2_o] - \eta^2\partial B^2_o/\partial u^2_o)$$

Proof. We "update" the control variable u^2 and then apply Corollary 1. Let us consider the system (19.8) for $k=\rho+1$ and with u^2 playing the role of u. In order to verify that all hypotheses of Corollary 1 are met by the $(n+k)$-dimensional vectors $\hat{B}^i_\nu$, $i=1,2$, one simply has to make use of the subsequent relations (cf. Lemma 19.3):

$$[\hat{B}^1_\nu,\hat{B}^1_\mu] = ([B^1_\nu,B^1_\mu]^T,0)^T,\ [\hat{B}^2_\nu,\hat{B}^2_\mu] = 0 \quad \text{if} \quad \nu,\mu\leq\rho,$$

$$[\hat{B}^1_\nu,\hat{B}^2_\mu]\in\hat{\mathcal{L}} \text{ if } \nu,\mu\leq\rho.$$

The statement in the second line follows from the hypothesis and from Lemma 13.2. In order to demonstrate that the assertion in question follows from the conclusion of Corollary 1 note that

$$[\hat{B}^2_\rho,\hat{B}^2_{\rho+1}] = -\partial B^2_o/u^2_o \quad \text{and}$$

$$[\hat{B}^1_\rho,\hat{B}^2_{\rho+1}] + [\hat{B}^2_\rho,\hat{B}^1_{\rho+1}] = (p^T,0)^T,\quad p=(-1)^{\rho+1}[B^1_\rho,B^2_o] + (-1)^\rho\partial B^1_{\rho+1}/\partial u^2_o .$$

Assume now that ρ is odd. Then $p = [B^1_\rho,B^2_o] - \partial B^1_{\rho+1}/\partial u^2_o$. On the other hand, from (21.2) and the hypothesis of the corollary on sees that

$[B^1_\rho,B^2_o] = -\partial B^1_{\rho+1}/\partial u^2_o + b, b\in\mathcal{L}$ and this gives the desired result.

By updating u^1_o one can pass from an even ρ to $\rho+1$.

In order to illustrate possible applications of the foregoing results we resume the discussion of the example (20.14). Now we will really exploit our assumption that Θ is unrestricted (this was actually not used in Sec. 20, hence our previous results concerning the equation (20.14) would in fact hold true also if Θ would be subject to constraints). If T satisfies the condition $0<T<T_{Max}$ for all $t\in I$, then we have indeed a "doubly singular arc" and we wish to find out whether one can gain additional information from the necessary conditions obtained in this chapter. We identify Θ with the first and T with the second component of the control variable. For the reader's convenience we have listed below in the first line the symbols used throughout this section and in the second line their respective meaning for the example in question.

(21.21)

$u^1=u^1_o$	$\dot u^1=u^1_1$	$u^2=u^2_o$	B^1_o
Θ	$\dot\Theta$	T	$(0,(T/m)Q'(\Theta),\ 0)^T$

$B^2_o(=b_o)$	$B^2_1(=b_1)$
$(0,m^{-1}Q(\Theta),-c^{-1})^T$	$m^{-1}(-Q(\Theta),Q'(\Theta)\dot\Theta,0)^T$

(' denotes differentiation with respect to Θ. The symbols b_o,b_1 refer to the notation used earlier cf. (20.14)-(20.16)). One easily verifies that the following identities hold true (cf. also (20.18))

$$\partial B^2_o/\partial u^2_o = 0,\ \partial B^2_o/\partial u^1_o = \frac{1}{T}B^1_o,\ [B^2_o,B^2_1] = \frac{1}{cm}B^2_1 .$$

If the vectors appearing in this line are evaluated along the singular arc allturn out to be elements of the linear subspace $\mathcal{L}(t)$ and this means that the hypotheses of Theorem 21.1 are satisfied. Hence for the singular arc the adjoint state vector y has to be orthogonal both to $[B^2_o,B^1_o]$ and to $\partial B^2_1/\partial u^1_o$. One finds now from (21.21) that

$$[B^2_o,\ B^1_o] = \frac{1}{cm}B^1_o\ ,$$

(21.22)

$$\partial B^2_1/\partial u^1_o = \frac{1}{m}(-Q'(\Theta),\ Q''(\Theta)\dot\Theta,0)^T.$$

The interesting feature of the second formula is the fact that it involves Θ,y only. Hence we have t w o independent equality type necessary conditions which relate Θ and y along a singular arc, namely

$$y^T B_1^2 = 0, \quad y^T(\partial B_1^2/\partial u_o^1) = 0 \ .$$

One can write these conditions in a more transparent way if the adjoint state variable y is partitioned according to the representation of the state variable (cf. Sec. 20), say as $(y_1, y_2, y_3)^T$. Using this notation the above relations can be brought into the form

$$-y_1^T Q(\theta) + y_2^T Q'(\theta)\dot{\theta} = 0, \quad -y_1^T Q'(\theta) + y_2^T Q''(\theta)\dot{\theta} = 0 \ .$$

Note that the first relation is a standard first order condition. The second condition can in fact also be derived from first order necessary conditions though this is not so obvious. One has to use the expression for $[B_o^2, B_o^1]$ (cf. (21.22)) and the identity

$$[B_o^2, B_o^1] = \partial B_1^2/\partial u_o^1 - \Gamma(\partial B_o^2/\partial u_o^1)$$

which is a special case of (21.2). It would be of interest to find out to which extent the necessary conditions obtained in this section so far are actually disguised first order conditions. This is certainly true if a normality condition holds, i.e. if $\mathcal{L}(t)$ has maximal rank and hence is the orthogonal complement of the adjoint vector. However an assumption of this kind was not necessary for the foregoing results, this seems to be an advantage of our approach.

PART III: FREE-ENDPOINT PROBLEMS.

For optimal control problems without terminal constraints a slightly improved version of the previous results holds. Roughly speaking, the linear space $\mathcal{L}(t)$ can be replaced by the orthogonal complement of the adjoint state variable $y(t)$. This will be made precise in Sec.22. In addition to these results there is a genuine s e c o n d order condition which so far has no analogue for constrained-endpoint problems. This condition will be presented in Sec. 23, it is a generalization of a condition due to Jacobson and Gabasov.

22. Restatement of previous results.

Let $u(\cdot),x(\cdot)$ be a fixed reference pair and $y(\cdot)$ a solution of the adjoint variational equation $\dot{y}= -f_x(t,x(t);u(t))y$. We denote by $\hat{\mathcal{L}}_{y(\cdot)}(t)$ the orthogonal complement of $y(t)$, i.e. the linear space $\{z:y(t)^T z = 0\}$. It follows then from the definition of the operator Γ (cf. Sec. 13) that the relation $g(t,x(t),\mathbf{U}(t))\in \hat{\mathcal{L}}_{y(\cdot)}(t)$ for all t in an open set $\mathcal{S}$ implies $\Gamma(g))(t,x(t),\mathbf{U}(t))\in\hat{\mathcal{L}}_{y(\cdot)}(t)$ for all $t\in\mathcal{S}$. Hence Lemma 19.1 remains valid - without any assumption about the dimension of $\hat{\mathcal{L}}_{y(\cdot)}(t)$ - if $\mathcal{L}(t)$ is replaced by $\hat{\mathcal{L}}_{y(\cdot)}(t)$.

Let the reference pair be now such that it solves the variational problem

$$\text{Minimize } \varphi(x(t_e)) \text{ subject to the constraints}$$

$$\dot{x} = f(t,x;u),\ u\in U,\ x(t_o) = x_o. \tag{22.1}$$

Here $\varphi(\cdot)$ denotes a scalar function of x, which is of class C^∞ . Let $y(\cdot)$ be the solution of the adjoint variational equation satisfying the initial condition

$$y(t_e) = \varphi_x(x(t_e)) . \tag{22.2}$$

It follows then from Theorem 9.1 that the inequalities $-y(t)^T p\leq 0$ *) hold for all $p\in \Pi_t$ and all $t\in[t_o,t_e]$. However the absence of terminal constraints allows to establish a stronger version of the higher order necessary condition than the one which we used up to now. This version is expressed by the subsequent theorem and its corollary.

*) Note that $-y(\cdot)$ is the multiplier to which the theorem refers, since it satisfies the correct transversality condition. Therefore signs are reversed throughout this section.

<u>Theorem 22.1.</u> Let $u(\cdot)$, $x(\cdot)$, $y(\cdot)$ have all properties mentioned above and let there be given a control variation concentrated at t with the corresponding trajectory $x(\cdot;\lambda,a)$ having the asymptotic expansion

$$(22.3)\quad x(t;\lambda,x_o) \approx x(t) + \lambda^r \sum_{\nu=o}^{\infty} \lambda^\nu p_\nu$$

(for the definition of a control variation and the definition of $x(\cdot;\lambda,a)$ see (9.3) and the subsequent remarks). Then the first non-vanishing among the real numbers $y(t)^T p_\nu$, $\nu=0,1,\ldots\, r-1$, is positive.

<u>Proof.</u> Let $\Phi(t,\tau)$ be the transition matrix of the variational equation (cf. (10.1)). We have then, in view of (22.3)

$$x(t_e;\lambda,x_o) = x(t_e) + \lambda^r \sum_{\nu=0}^{r-1} \lambda^\nu \Phi(t_e,t)p_\nu + \mathcal{O}(\lambda^{2r})$$

and

$$\varphi(x(t_e;\lambda,x_o) = \varphi(x(t_e)) + \varphi_x(x(t_e))^T\{\lambda^r \sum_{\nu=0}^{r-1} \lambda^\nu \Phi(t_e,t)p_\nu\} + \mathcal{O}(\lambda^{2r}) \;.$$

On the other hand $\varphi(x(t_e;\lambda,x_o)) \geq \varphi(x(t_e))$ for $0 \leq \lambda \leq \lambda_o$ since $x(t_e;\lambda,x_o)$ is the terminal point of an admissible trajectory initiating at x_o. The above relation can now be written in the form

$$\varphi(x(t_e;\lambda,x_o)) = \varphi(x(t_e)) + \lambda^r \sum_{\nu=0}^{r-1} \lambda^\nu y(t)^T p_\nu + \mathcal{O}(\lambda^{2r})$$

and the conclusion of the theorem becomes evident.

<u>Corollary.</u> Let there be given a positive integer N, scalar functions $z_1(\lambda), \ldots, z_N(\lambda)$ and sequences $\mathbf{U}_i(\lambda) = \{u_{i,o}(\lambda),\ldots\}$, $i=1,\ldots,N$, such that the conditions (i) - (iii) of Definition 8.1 are satisfied. for some fixed t and $x=x(t)$. Let $\mathbf{U}(t)$ be the sequence associated with the reference control $u(\cdot)$ according to (7.5).

We put

$$(22.4)\quad C^*(\lambda) := C(t,x(t),\, z_1(\lambda),\ldots,z_N(\lambda);\, \mathbf{U}(t),\, \mathbf{U}_1(\lambda),\ldots,\mathbf{U}_N(\lambda)) - x(t)$$

$C^*(\lambda)$ is then a formal power series in λ without constant terms, i.e. $C^*(\lambda)$ admits an expansion of the form

$$C^*(\lambda) = \lambda^r \sum_{\nu=o}^{\infty} \lambda^\nu p_\nu \;.$$

Conclusion: The first non-vanishing among the numbers $y(t)^T p_\nu$, $\nu=0,\ldots,r-1$, - if there are any - is positive.

Proof. That $C(t,x(t),z_1(\lambda),\dots,z_N(\lambda);\,\mathcal{U}(t),\,\mathcal{U}_1(\lambda),\dots,\mathcal{U}_N(\lambda))$ is a formal power series in λ follows from the considerations preceding Definition 9.1. On the other hand this power series represents the asymptotic expansion of $x(t;\lambda,x_o)$ for a suitable control variation concentrated at t (see remark 3 in connection with Definition 9.2). This proves the corollary.

We now will apply the corollary to the situation considered in Sec.20 and state a version of the generalized Clebsch-Legendre condition which takes the present character of the reference solution into account. $v(t) = (v^1(t),\dots,v^m(t))^T$ and $b_\nu(t,x)$ have the same meaning as in Sec. 20 (cf. (20.9)).

Theorem 22.2. Let the hypotheses concerning $u(\cdot),x(\cdot),y(\cdot)$ be the same as in Theorem 22.1. Assume in addition that we have

$$(22.5)\quad \begin{aligned} &\text{(i)}\quad u(t)+\xi v(t)\in U \qquad \text{for } |\xi| \text{ sufficiently small,}\\ &\text{(ii)}\quad \sum_{i,j=1}^{m} H_{u^i u^j}(t,x(t),y(t),u(t))v^i(t)v^j(t) = 0 \end{aligned}$$

for all $t\in I$ where I is a closed subinterval of $[t_o,t_e]$ and

$$H(t,x,y,u) = y^T f(t,x;u).$$

Then the first among the functions $y(t)^T[b_{\nu-1},b_\nu](t,x(t))$, $\nu=1,2,\dots$, which does not vanish identically is non-positive on I.

Proof. For each fixed t the scalar function $H(t,x(t),y(t),u(t)+\xi v(t))$ has a relative minimum at $\xi=0$. This follows from the maximum principle and hypothesis (i). Hence we have $y(t)^T b_o(t,x(t))=0$ for all t and therefore - see the remarks at the beginning of this section - also

$$(22.6)\quad y(t)^T b_\nu(t,x(t)) = 0 \quad \text{for all } t \text{ and all } \nu\geq 0.$$

The proof amounts then more or less to a repetition of the proofs given in Sec. 20. We have $b_\nu(t,x) = \tilde{b}_\nu(t,x,0)$, where $\tilde{b}_\nu$ is defined as in Sec. 20. (See (20.1)). Using the notation introduced recently the statement of the theorem can then be phrased as follows. If we have

$$(22.7)\quad \frac{\partial\tilde{b}_o}{\partial\xi}(t,x(t),0)\in\hat{\mathcal{L}}_{y(\cdot)}(t),\quad [\tilde{b}_{\nu-1},\tilde{b}_\nu](t,x(t),0)\in\hat{\mathcal{L}}_{y(\cdot)}(t)$$

for $\nu=1,\dots,\rho$ and all $t\in I$, then

$$y(t)^T\,[\tilde{b}_\rho,\tilde{b}_{\rho+1}](t,x(t),0)\leq 0.$$

Again we may assume for the purpose of the proof that the relations (20.4) hold true. Note that the transformation of the control variable which we used to establish (20.4) does not involve the state variable. Therefore the variational equation and also the adjoint state vector $y(\cdot)$ remain unchanged. (22.6) and (22.7) can then be rewritten in the form

$$(22.8)\quad \begin{aligned} &B^1_\nu(t,x(t),\mathcal{U}(t)) \in \hat{\mathcal{L}}_{y(\cdot)}(t), \quad \nu = 0,1,\ldots, \\ &(\partial B^1_o/\partial u^{(1)}_o)(t,x(t),\mathcal{U}(t)) \in \hat{\mathcal{L}}_{y(\cdot)}(t), \\ &[B^1_{\nu-1},B^1_\nu](t,x(t),\mathcal{U}(t)) \in \hat{\mathcal{L}}_{y(\cdot)}(t), \quad \nu=1,\ldots,\rho . \end{aligned}$$

It follows now from what was said at the beginning of this section that the statement of Lemma 19.1 and the statements (20.8) remain true if $\mathcal{L}(t)$ is replaced by the space $\hat{\mathcal{L}}_{y(\cdot)}(t)$. One can then proceed as in the proof of Lemma 18.3 and construct a formal power series $\tilde{C}(\lambda)$ (with $\mu=2\rho+2$) to which the corollary applies and which satisfies (18.16) except for the fact that the b_ν now belong to the space $\hat{\mathcal{L}}_{y(\cdot)}$. The conclusion is then $(-1)^{\rho+1}y(t)^T(\partial B^1_{2\rho+2}/\partial u^{(1)}_o)(t,x(t),\mathcal{U}(t)) \geq 0$, and this relation can be brought into the form as stated in the theorem, again by making use of (20.8) (with $\mathcal{L}$ replaced by $\hat{\mathcal{L}}$.

By the same type of argument one can establish in the case of multivariable controls results which differ from those stated in Sec. 21 simply by the fact that $\hat{\mathcal{L}}_{y(\cdot)}$ plays in the role of $\mathcal{L}$.

23. The Jacobson-Gabasov condition and its generalizations.

One observes from the explicit formulas given in Lemma 18.2 that the coefficients of the formal power series $\tilde{C}(\lambda)$ are linear combinations of three types of quantities, namely the derivatives $\partial B^1_\tau/\partial u^1_\kappa$, the Lie-brackets $[B^1_\nu,B^1_\mu]$ and finally the "half-Lie-brackets" $(B^1_\mu)_x B^1_\nu$. The latter ones have so far played no role. The reason will become apparent if one analyses the approach which led to Theorem 20.1. In order to apply Theorem 9.1 it was necessary to specialize the parameters in the formal power series $\tilde{C}(\lambda)$ in such a way that all relevant lower order terms involving the half-Lie-brackets disappear. If Theorem 22.1 however becomes applicable one has more freedom in the choice of the parameters. Hence there exists for problems without terminal constraints a further type of necessary condition which is based on the products $(B^1_\mu)_x B_\nu$ itself (and not on their skew-symmetric parts). A special case of this criterion can be found in the literature (see e.g. [7] , Sec. 3.3 and [11] , Ch. VI, §5, Theorem 13). The general case will be stated below as Theorem 23.1. We assume from now on that u is scalar, hence $B_\nu = B_\nu^1$ is a column vector. The extension of our result to systems of arbitrary control dimension m can be obtained simply by applying the criterion to the system $\dot{x}=\tilde{f}(t,x,\xi)$, where $\tilde{f}$ is defined as in Sec. 20 with ξ playing then the role of the control variable.

Lemma 23.1 Let $x(t;t_o,a)$ be the general solution of some differential eq. $\dot{x} = f(t,x)$ and let $\tilde{x}(t)$ be a particular solution which exists on some interval $[t_o,t_e]$. If $\varphi(x)$ is a sufficiently smooth scalar function of x then we have an asymptotic relation of the form

$$\begin{aligned}\varphi(x(t_e;t_o,a)) &= \varphi(\tilde{x}(t_e)) + (a-\tilde{x}(t_o))^T y(t_o) + \\ &+ \frac{1}{2}(a-\tilde{x}(t_o))^T P(t_o)(a-\tilde{x}(t_o)) + \mathcal{O}(\|a-\tilde{x}(t_o)\|^3) .\end{aligned} \tag{23.8}$$

Here $y(\cdot)$, $P(\cdot)$ respectively is a n-dimensional vector and a symmetric (n,n)-matrix respectively which are defined in terms of the following linear initial value problem

$$\begin{aligned}&y(t_e) = \varphi_x(\tilde{x}(t_e)),\quad P(t_e) = (\varphi_{x^i x^j}(\tilde{x}(t_e)))\ (=\text{Hessian of } \varphi\ ),\\ &\dot{y} = -A(t)^T y\ ,\quad \dot{P} = -A(t)^T P - PA(t) - K(t,y(t)),\end{aligned} \tag{23.9}$$

where $A(\cdot),K(\cdot)$ are matrices of type (n,n) and given in explicit terms as

$$A(t) = f_x(t,\tilde{x}(t)),\quad K(t,y) = (y^T f_{x^i x^j}(t,\tilde{x}(t))) . \tag{23.10}$$

Proof. Let $\Phi(t,\tau)$ be the transition matrix of the linear variational equation $\dot{y} = A(t)y$. One observes then that $y(\cdot),P(\cdot)$ can be written in the following form

$$\begin{aligned}&y(t) = \Phi(t_e,t)^T \varphi_x(\tilde{x}(t_e)) ,\\ &P(t) = \int_t^{t_e} \Phi(\tau,t)^T K(\tau,y(\tau))\Phi(\tau,t)d\tau + \Phi(t_e,t)^T P_e \Phi(t_e,t),\end{aligned} \tag{23.11}$$

where

$$P_e = (\varphi_{x^i x^j}(\tilde{x}(t_e))) . \tag{23.12}$$

We are now going to establish the following result. Let b be a fixed but arbitrary vector, λ a parameter and let $x(t,\lambda)$ be the solution of the initial value problem

$$\dot{x} = f(t,x),\qquad x(t_o) = \tilde{x}(t_o) + \lambda b. \tag{23.13}$$

Then we have these relations

$$(23.14)\quad \begin{aligned} b^T y(t_o) &= \varphi_x(\tilde{x}(t_e))^T x_\lambda(t_e,0) \; , \\ b^T P(t_o) b &= \varphi_x(\tilde{x}(t_e))^T x_{\lambda\lambda}(t_e,0) + x_\lambda(t_e,0)^T P_e x_\lambda(t_e,0) \; . \end{aligned}$$

Once this result has been proved it is clear that we have the asymptotic representation

$$(23.15)\quad \varphi(\tilde{x}(t_e)) + \lambda b^T y(t_o) + \frac{1}{2}\lambda^2 b^T P(t_o) b + \mathcal{O}(\lambda^3)$$

for the function $\varphi(x(t_e,\lambda))$. On the other hand the solution of the initial value problem (23.13) can be written in the form $x(t;t_o,\tilde{x}(t_o)+\lambda b)$. It is then clear, by standard arguments concerning Taylor-expansions of functions of several variables, that the desired result (23.8) is a consequence of (23.15).

In order to prove (23.14) we observe that $x_\lambda(t,0)$ and $x_{\lambda\lambda}(t,0)$ respectively are solutions of the following initial value problems respectively

$$\dot{y} = A(t)y \; , \quad y(t_o) = b,$$

$$\dot{y} = A(t)y + \sum_{i,j} f_{x^i x^j}(t,\tilde{x}(t))\, x_\lambda^i(t,0)\, x_\lambda^j(t,0) \, , \quad y(t_o)=0 \; .$$

Hence we have, from the variations of constants formula

$$(23.16)\quad \begin{aligned} x_\lambda(t,0) &= \Phi(t,t_o)b, \\ x_{\lambda\lambda}(t,0) &= \int_{t_o}^{t} \Phi(t,\tau)\Big(\sum_{i,j} f_{x^i x^j} x_\lambda^i x_\lambda^j\Big)d\tau \; . \end{aligned}$$

It follows then from 23.11 that

$$x_\lambda(t_e,0)^T \varphi_x(\tilde{x}(t_e)) = b^T y(t_o),$$

and this is the first of the relations (23.14). Also (cf. (23.11), (23.16))

$$\begin{aligned} \varphi_x(\tilde{x}(t_e))^T x_{\lambda\lambda}(t_e,0) &= \int_{t_o}^{t_e} y(\tau)^T\Big(\sum_{i,j} f_{x^i x^j}(\tau,\tilde{x}(\tau))\, x_\lambda^i(\tau,0)\, x_\lambda^j(\tau,0)\Big)d\tau \\ &= b^T\Big(\int_{t_o}^{t_e} \Phi(\tau,t_o)^T K(\tau,y(\tau))\, \Phi(\tau,t_o)\, d\tau\Big) b \; . \end{aligned}$$

Finally, from (23.16)

$$x_\lambda(t_e,0)^T P_e x_\lambda(t_e,0) = b^T \Phi(t_e,t_o)^T P_e \Phi(t_e,t_o) b \; .$$

Combining the two last relations with (23.11) yields the second line of (23.14). Thereby the lemma is proved.

Corollary. Given a control system of the form (7.1) and let $\tilde{u}(\cdot),\tilde{x}(\cdot)$ again be a reference pair which satisfies condition (22.1). Let, for some $\tilde{t}\in[t_o,t_e]$, $C^*(\lambda)$ be a formal power series of the type (22.4) (with $t=\tilde{t}$) and let $C^*(\lambda) = \mathcal{O}(\lambda^r)$. Consider then the formal power series

$$(23.17) \qquad y(\tilde{t})^T C^*(\lambda) + \tfrac{1}{2}(C^*(\lambda))^T P(\tilde{t}) C^*(\lambda) \quad {}^{1)}.$$

Claim: If (23.17) is written in explicit terms as $\lambda^r \sum_{\nu=0}^{\infty} \lambda^\nu \pi_\nu$ then the first non-vanishing among the numbers π_ν, $\nu=0,\dots,2r-1$ - if there is any - is positive.

Proof. There exists a control variation $u(\cdot,\lambda)$ concentrated at $\tilde{t}$ such that the solution $\tilde{x}(t,\lambda)$ of the initial value problem

$$\dot{x} = f(t,x;u(t,\lambda)), \quad x(t_o) = \tilde{x}(t_o) ,$$

is connected with the given formal power series through the relation

$$\tilde{x}(\tilde{t},\lambda) - \tilde{x}(\tilde{t}) \approx C^*(\lambda) .$$

On the other hand $\tilde{x}(t,\lambda)$ is for $t\geq\tilde{t}$ solution of the initial value problem

$$\dot{x} = f(t,x;\tilde{u}(t)), \quad x(\tilde{t}) = \tilde{x}(\tilde{t},\lambda) .$$

If the lemma is then applied with $\tilde{t},\tilde{x}(\tilde{t},\lambda)$ respectively playing the roles of t_o, a respectively one sees that the expression (23.17) represents $\varphi(\tilde{x}(t_e,\lambda))$ up to terms of order $\mathcal{O}(\lambda^{3r})$. On the other hand, since the reference pair is supposed to minimize $\varphi(x(t_e))$, we have $\varphi(\tilde{x}(t_e,\lambda))\geq\varphi(\tilde{x}(t_e,0)) = \varphi(\tilde{x}(t_e))$ for $\lambda\geq 0$. The statement of the lemma is now evident.

Before we state our main result we remind the reader that we deal here with systems which are of control dimension 1, hence u is scalar and $B_\nu = B_\nu^1$ is a column vector.

Theorem 23.1 Given $u(\cdot),x(\cdot),y(\cdot)$ with all properties as stated in Theorem 22.2. Let the sequence $\mathbf{U}(t)$ and the matrices $A(t),K(t),P(t)$ respectively be defined according to (7.5) and (23.9), (23.10) respectively. Furthermore assume that these relations hold, for some integer $\rho\geq0$ and for all t in some open subset $\mathcal{S}$ of $[t_o,t_e]$ (the last one is meaningless for $\rho=0$) : $u(t)\in$ int U and

$$(23.18) \qquad \begin{aligned} &(\partial B_o/\partial u_o)(t,x(t),\mathbf{U}(t))\in \hat{\mathcal{L}}_{y(\cdot)}(t) , \\ &[B_{\nu-1},B_\nu](t,x(t),\mathbf{U}(t))\in \hat{\mathcal{L}}_{y(\cdot)}(t), \quad \nu=1,\dots,\rho . \end{aligned}$$

---

1) For the definition of $y(\cdot)$, $P(\cdot)$ use (23.9) and (23.10) and identify $f(t,x)$ with $f(t,x;\tilde{u}(t))$.

Consider then the quadratic form in variables $\xi_o,\ldots,\xi_\rho$ which is given as (for the definition of y,P cf. the corollary to Lemma 23.1).

$$(23.19\quad \sum_{\mu,\nu=0}^{\rho} \xi_\mu\xi_\nu\, y^T(B_\mu)_x B_\nu + (\sum_{\mu=0}^{\rho} \xi_\mu B_\mu^T)P(\sum_{\mu=0}^{\rho} \xi_\mu B_\mu)$$

where $y=y(t)$, $P=P(t)$, $B_\nu=B_\nu(t,x(t),\mathbf{U}(t))$, $(B_\mu)_x=(B_\mu)_x(t,x(t),\mathbf{U}(t))$.
Claim: For every $t\in\mathcal{S}$ (23.19) is a positive semi-definite form.

Remarks. 1.) In case $\rho=0$ the statement can be expressed in terms of the Hamiltonian $H(t,x,y,u) = y^T f(t,x,u)$ and then turns out to be identical with Jacobson's condition, namely: If $u\in\mathrm{int}U$ and $H_u=0$ for $t\in\mathcal{S}$ then $y^T(B_o)_x B_o + B_o^T P B_o = H_{ux}^T f_u + f_u^T P f_u \geq 0$ (cf. the reference above).
2) We have $y^T(B_\mu)_x B_\nu = y^T(B_\nu)_x B_\mu$ if $\nu,\mu\leq\rho$, hence the coefficients which appear in the first part of (23.19) are symmetric functions of μ,ν, as they should be. This follows from these relations

$$(23.20)\quad \begin{aligned} &[B_\nu,B_\mu](t,x(t),\mathbf{U}(t))\in\hat{\mathcal{L}}_{y(\cdot)}(t) \quad \text{if}\quad \nu,\mu\leq\rho\ . \\ &(\partial B_\lambda/\partial u_o)(t,x(t),\mathbf{U}(t))\in\hat{\mathcal{L}}_{y(\cdot)}(t)\quad \text{if } \lambda\leq 2\rho+1, \end{aligned}$$

which hold for all $t\in\mathcal{S}$. That (23.20) is correct can be inferred from the hypothesis (23.18) by the same argument as we used earlier (cf. (22.8) and the subsequent remarks , cf. also Theorem 17.2).

Proof of Theorem 23.1. We take $t\in\mathcal{S}$ fixed and write $\tilde{x}$, $\hat{\mathcal{L}}$ respectively instead of $\tilde{x}(t)$, $\hat{\mathcal{L}}_{y(\cdot)}(t)$ respectively. Let r be a positive integer such that

$$(23.21)\quad r+\rho+1<2r+1,\quad 2r+2\rho+3\leq 3r+1\ .$$

Given now $\rho+1$ real numbers $\xi_o,\ldots,\xi_\rho$. We introduce a real parameter λ and choose r real numbers $z_1,\ldots,z_r$ and sequences $\mathbf{V}_o=\mathbf{O}$, $\mathbf{V}_1,\ldots,\mathbf{V}_r$ such that we have identically in x

$$(23.22)\quad \frac{1}{\nu!}\sum_{i=1}^{r} l^{(\nu)}(t,x;\mathbf{U}_o,\mathbf{U}_{i-1}\mathbf{U}_i)(\lambda z_i)^\nu = \begin{cases} 0 & \text{if}\quad \nu<\rho+1\ , \\ \lambda^{r+\rho+1}p(x) & \text{if}\quad \nu=\rho+1\ . \end{cases}$$

where

$$(23.23)\quad \begin{aligned} &\mathbf{U}_o=\mathbf{U}(t) = \{u(t),\dot{u}(t),\ldots\}\ ,\quad \mathbf{U}_i=\mathbf{U}_o + \lambda^r\mathbf{V}_i,\ i=1,\ldots,r, \\ &p(x) = \sum_{\mu=o}^{\rho}\xi_\mu B_\mu(t,x). \end{aligned}$$

That such a choice of $\mathcal{V}_i$ is possible can be seen by the same type of argument as used in the proof of Lemma 14.2 (one simply has to construct a formal power series of the type considered there with $c_{\nu,\mu} = 0$ for $\nu<\rho+1$ and $c_{\nu,\mu} = \xi_\mu$ for $\nu=\rho+1$). Furthermore it follows from the definition of $1^{(\nu)^{\nu,\mu}}, L^{(\nu)^\mu}$ (cf. Sec. 14) and from (23.20) that for $x=\tilde{x}=\tilde{x}(t)$ these relations hold

$$(23.24)\quad \begin{aligned} &[1^{(\nu)}(\mathcal{U}_o, \mathcal{U}_{j-1}-\mathcal{U}_j), 1^{(\mu)}(\mathcal{U}_o, \mathcal{U}_{i-1}-\mathcal{U}_i)] \in \hat{\mathcal{L}} \text{ if } \nu+\mu\leq 2\rho+2 \\ &L^{(\nu)}(\mathcal{U}_o, \mathcal{U}_{i-1}-\mathcal{U}_i)\cdot(\mathcal{U}_{i-1}+\mathcal{U}_i-2\,\mathcal{U}_o) \in \hat{\mathcal{L}} \text{ if } \nu\leq 2\rho+2 \,. \end{aligned}$$

We now introduce the formal power series

$$C^*(\lambda) = C(t,\tilde{x},\lambda z_1,\ldots,\lambda z_r; \mathcal{U}_o, \mathcal{U}_1,\ldots, \mathcal{U}_r) - \tilde{x}$$

and the corresponding series $C^{(\nu)}(\lambda) = C^{(\nu)}(t,\tilde{x},\ldots), \check{C}(\lambda)$ (cf.Sec.14) where the $\mathcal{U}_i$ are given as in (23.23). It follows then from (23.21) - (23.24) that we have these representations of the series in question (see the explicit formulas (14.3),(14.6) for $C^{(2)}, \check{C}^{(2)}$):

$$\begin{aligned} C^*(\lambda) + \tilde{x} &= C^{(2)}(\lambda) + \mathcal{O}(\lambda^{3r+1}) \,, \\ C^{(1)}(\lambda)+\tilde{x} &= \lambda^{r+\rho+1}(p(\tilde{x}) + \sum_{\nu=1}^{\infty} \lambda^\nu b_\nu) \,, \\ C^{(2)}(\lambda) &= C^{(1)}(\lambda) + \lambda^{2r} \sum_{\nu=1}^{2\rho+2} \lambda^\nu b_\nu + \check{C}^{(2)}(\lambda) + \mathcal{O}(\lambda^{2r+2\rho+3}) \,, \\ 2\check{C}^{(2)}(\lambda) &= \lambda^{2(r+\rho+1)} p_x(\tilde{x})p(\tilde{x}) + \lambda^{2r} \sum_{\nu=2}^{2\rho+2} \lambda^\nu b_\nu + \mathcal{O}(\lambda^{2r+2\rho+3}) \,, \end{aligned}$$

where b_ν denotes an element of $\hat{\mathcal{L}}$ (not necessarily always the same). Since the adjoint vector $y(t)$ is orthogonal to the elements of $\hat{\mathcal{L}}$ we are finally arrived at these relations (cf. (23.21))

$$\begin{aligned} C^*(\lambda) &= \lambda^{r+\rho+1} p(\tilde{x}) + \mathcal{O}(\lambda^{r+\rho+2}) \,, \\ 2y(t)^T C^*(\lambda) &= \lambda^{2(r+\rho+1)} y(t)^T p_x(\tilde{x}) p(\tilde{x}) + \mathcal{O}(\lambda^{2(r+\rho+1)+1}) \,. \end{aligned}$$

The conclusion of the theorem follows then immediately from the corollary of Lemma 23.1.

If the underlying system is linear in the control variable u it is clear that the hypothesis of the theorem is always satisfied if we take $\rho=0$ and it is satisfied for $\rho=1$ provided $y(\cdot)$ is orthogonal to the Lie-bracket $[B_o,B_1]$. A more detailed analysis of the proof of the preceding theorem shows that the last condition need not be satisfied along the whole arc - as stated in the theorem - in order to draw the conclusion for $\rho=1$. This will be made more precise in the subsequent corollary.

Corollary 1. Let the control system be linear in the (scalar) control variable u. Assume that $u(t)\in \text{int}U$ for all $t\in \mathcal{S}$ and that

(23.25) $\quad y(\tilde{t})^T([B_o,B_1](\tilde{t},x(\tilde{t})))= 0$

holds for some $\tilde{t}\in \mathcal{S}$. Furthermore assume that the reference control is continuous at $t=\tilde{t}$. Consider then the quadratic form (23.19) for $\rho=1$, where y,B_ν,P have to be understood as the values of these quantities for $t=\tilde{t}$ and $x=x(\tilde{t})$. Claim: This form is positive semi-definite.

Proof. We first make use of the generalized Clebsch-Legendre condition (Theorem 20.2). Since the underlying system is linear in u the condition (20.10) is satisfied and hence the inequality (20.13') holds for all $t\in \mathcal{S}$ if $b_\rho(t,x)$ is identified with $B_o(t,x)$. In other words, the scalar function of t which is given as

$$-y(t)^T([B_o,B_1](t,x(t))$$

is non-negative and hence assumes a relative minimum at $t=\tilde{t}$ if (23.25) holds true. Now the continuity of the reference control implies differentiability at $t=\tilde{t}$ of $y(\cdot)$, $x(\cdot)$ and hence also of the above function. If the derivative is calculated according to (13.2) one arrives at a further relation namely

(23.26) $\quad y(\tilde{t})^T((\Gamma[B_o,B_1])(\tilde{t},x(\tilde{t}),\mathbf{u}(\tilde{t}))) = 0.$

The proof is then easily completed with the help of Theorem 17.2. Since the system is supposed to be linear in u the derivatives $\partial B_\lambda/\partial u$ vanish identically in t,x for $\lambda=0,1$ and one obtains from the formula of the theorem for $\sigma=\rho,\mu= 1,\nu= 0,1$ these identities

$$[B_1,B_o] = -\partial B_2/\partial u, \quad 0 = [B_1,B_1] = -\partial B_3/\partial u + 2\Gamma(\partial B_2/\partial u) .$$

Both combined yield a further identity, namely

$$\partial B_3/\partial u = 2\Gamma([B_o,B_1])$$

It follows then from these relations and from (23.25), (23.26) that for $t=\tilde{t}$ the adjoint variable is orthogonal to $\partial B_\lambda/\partial u$ for $\lambda\leq 3$.

In other words, the conditions (23.20) are satisfied for $\rho=1$ and $t=\tilde{t}$. The conclusion of Theorem 23.1 however is drawn from (23.20) only (note that the hypotheses of the theorem are used for the proof only in order to ensure the validity of (23.20)). Thereby it is clear that the statement of the corollary holds true.

Next we wish to discuss a special case of Theorem 23.1 where the statement involves the adjoint variable only.

Corollary 2. Assume that the right hand side f of the system equation is linear in the state variable x and assume that also the function φ is linear. Then the conclusion of Theorem 23.1 runs as follows: The quadratic form

$$\sum_{\mu,\nu=0}^{\rho} \xi_\mu \xi_\nu y^T (B_\mu)_x B_\nu \tag{23.19'}$$

is positive semi-definite.

Proof. Since f is linear in x the matrix $K(t,y)$ vanishes identically in y and for every t (cf. (23.10)). Hence the matrix differential equation for P is a homogeneous linear differential equation. Furthermore the value of P at $t=t_e$ is zero, since φ is a linear function. All this can be recognized immediately from the relations (23.9). Hence we have $P(t) = 0$ for all t and the quadratic form (23.19) reduces to the form as stated in the corollary.
The last corollary is useful if one wants to write down in a concrete case a further system of necessary conditions which are related to Theorem 23.1. To be more precise these conditions are consequences of the theorem in case the quadratic form (23.19) is not strictly positive (they are meaningless if the form is positive definite). The general principle from which these conditions can be derived will be described in the next theorem. In order to avoid notational confusion we denote for the remaining portion of this section the reference pair by $\tilde{u}(\cdot),\tilde{x}(\cdot)$ and the matrix which enters into the statement of Theorem 23.1 by $\tilde{P}(\cdot)$ (and not by $P(\cdot)$). In other words, $\tilde{P}(\cdot)$ is the solution of the initial value problem

$$\dot{P} = -A(t)^T P - PA(t) - K(t,y(t)), \quad P(t_e) = (\varphi_{x^i x^j}(\tilde{x}(t_e))) \tag{23.27}$$

(for the definition of $A(\cdot),K(\cdot)$ cf. (23.9), (23.10)).

Theorem 23.2 Let the underlying system be linear in the control variable and let all hypotheses of Theorem 22.2, 23.1 (or of Corollary 1) be satisfied for some $\rho\geq 0$. Assume that $\tilde{u}(t)\in \mathrm{int}U$ for all $t\in[t_o,t_e]$ and that the quadratic form (23.19) vanishes for some particular choice $t=\tilde{t}, \xi_\mu=\tilde{\xi}_\mu, \mu=0,\dots,\rho$ of the variables, where $\tilde{t}\in I$ and $\tilde{t}<t_e$. Put $\tilde{B}:=\sum_{\mu=0}^{\rho}\tilde{\xi}_\mu B_\mu(\tilde{t},x(\tilde{t}),\boldsymbol{U}(\tilde{t}))$. Then the matrix $\tilde{P}(\cdot)$-which is defined in terms of the initial value problem (23.27)- together with the n-dimensional column vector $v=0$ has the following property. The value of $\tilde{P}(\cdot)$ at $t=\tilde{t}$ minimizes the quantity $\tilde{B}^T P\tilde{B}$ in comparison with all pairs $v(\cdot)$, $P(\cdot)$ which satisfy these conditions:

$v(\cdot)$ is an n-dimensional column vector, the components are arbitrary piecewise C^∞-functions of t and vanish in a neighborhood of $\tilde{t}$. $P(\cdot)$ assumes the initial value $\tilde{P}(t_e)$ at $t=t_e$ and is solution of the matrix differential equation

$$(23.28)\qquad \begin{aligned}\dot{P} &= -A(t)^T P - PA(t) - K(t,y(t))\\ &\quad -(PB_o(t)+s(t))v(t)^T - v(t)(B_o(t)^T P+s(t)^T)\end{aligned}$$

on the interval $[\tilde{t},t_e]$. The column vectors B_o, s are defined as follows

$$(23.29)\qquad B_o(t) := B_o(t,x(t)),\ s(t):=(B_o)_x(t,x(t))^T y(t)\ .$$

Remark. Note that we have found another necessary condition which is expressed in terms of a new variational problem. The pair $(v(t)\equiv 0, \tilde{P}(t))$ yields a minimum to this problem if the reference pair $\tilde{u}(\cdot),\tilde{x}(\cdot)$ is a solution of the original problem and if in addition the quadratic form (23.19) fails to be strictly positive for some $\tilde{t}$. As to the nature of the new variational problem it is easy to see that it can be reformulated in such a way that it fits into the standard form (22.1) except for the fact that the time direction is reversed (t_e plays the role of the initial-, $\tilde{t}$ the role of the terminal time). The state variable is of dimension $\frac{1}{2}n(n+1)$, its components are the elements of the symmetric matrix P. The control variable v is of dimension n and can assume arbitrary values. The admissible specializations of v however are subject to a condition which so far has not appeared in this work: They have to vanish in some neighborhood of $\tilde{t}$. It is easy to convince oneself that this restriction is irrelevant as far as the application of any of our previous results to the variational problem in question is concerned. Finally we wish to point out that the side condition for this problem is given as a differential equation which is linear both in the state and the control variable. Furthermore the cost functional is a homogeneous linear function of the terminal value of the state variable.

<u>Proof of Theorem 23.2.</u> Given an arbitrary n dimensional vector $v(\cdot)$ with the properties as stated in the theorem. $v(\cdot)$ has to be regarded as fixed throughout the proof. Our aim is to establish the inequality

$$\tilde{B}^T \tilde{P}(\tilde{t}) \tilde{B} \leq \tilde{B}^T P(\tilde{t}) \tilde{B} \tag{23.30}$$

where $P(\cdot)$ is the solution of (23.28) with initial value $\tilde{P}(t_e)$ at $t=t_e$.

We choose an open subset $\hat{U} \subseteq U$ and an open subset X of the (t,x)-space such that the following two conditions are satisfied

(23.31)
(i) $\tilde{u}(\tilde{t}\pm 0) \in \hat{U}$ and $(t,\tilde{x}(t)) \in X$ if $t \in [t_o,t_e]$,

(ii) $u \in \hat{U}$, $(t,x) \in X$ and $t \in [t_o,t_e]$ implies
$u+v(t\pm 0)^T(x-\tilde{x}(t)) \in U$.

Note that the reference control assumes values in the interior of U for all $t \in [t_o,t_e]$, hence one certainly can find $\hat{U}$ and X with the desired properties.

Since f is a linear function of u it can be represented in the form $f(t,x;0)+uB_o(t,x)$. We now introduce a function $\hat{f}=\hat{f}(t,x;u)$ as follows

$$\hat{f}(t,x;u) := f(t,x;0)+(u+v(t)^T(x-\tilde{x}(t)))B_o(t,x), \tag{23.32}$$

and observe that there is an obvious relation between the given differential equation (7.1) and the modified equation

$$\dot{x} = \hat{f}(t,x;u), \tag{23.33}$$

nameley: If $\hat{u}(\cdot),x(\cdot)$ is a solution of (23.33) then the pair $u(\cdot)$, $x(\cdot)$ is a solution of (7.1) where $u(\cdot)$ is defined in explicit terms as

$$u(t):=\hat{u}(t)+v(t)^T(x(t)-\tilde{x}(t)).$$

Furthermore we have $u(t) \in U$ for all $t \in [t_o,t_e]$ (i.e. $u(\cdot)$ is admissible) provided the pair $(\hat{u}(\cdot),x(\cdot))$ satisfies the conditions

$$\hat{u}(t\pm 0) \in \hat{U}, \quad (t,x(t)) \in X \tag{23.34}$$

for all $t \in [t_o,t_e]$. This follows from (23.31) part(ii), whereas part (i) implies that (23.34) in particular holds true for the reference pair. Furthermore it is obvious that the reference pair is a solution of (23.33); it is then also a solution of the following variational problem: Minimize $\varphi(x(t_e))$ subject to the constraints

$$(23.35)\qquad \begin{aligned} &\dot{x} = \hat{f}(t,x;u), u\in\hat{U}, \quad x(t_o) = \tilde{x}(t_o), \\ &(t,x)\in X \quad \text{for all} \quad t\in[t_o,t_e]. \end{aligned}$$

This is a variational problem of the type (22.1) with an additional state constraint. However both X and $\hat{U}$ are open sets and it is easy to see that the necessary conditions which we obtained for problems without state constraints carry over without any change to the present situation. The desired inequality (23.30) will now follow from Theorem 23.1 if the necessary conditions are evaluated at the time $t=\tilde{t}$ and the reference pair is considered as solution of the variational problem (23.35). In order to put this into evidence, let us compute the quantities $\hat{y},\hat{A},\hat{B}_\nu,\hat{P}$ which enter into the statement of the theorem in case the right hand side of the underlying differential equation is not $f(t,x;u)$ but

$$\hat{f}(t,x;u) = f(t,x;u) + v(t)^T(x-\tilde{x}(t))B_o(t,x)$$

(cf. (23.32)). Taking partial derivatives with respect to the components of x on both sides of the above line and substituting the values of the reference pair for u and x one arrives at these relations (the omitted argument is $t,\tilde{x}(t),\tilde{u}(t)$):

$$(23.36)\qquad \begin{aligned} &\hat{f}_x(\ldots) = f_x(\ldots) + B_o(\ldots)v(t)^T, \\ &\hat{f}_{xx^i}(\ldots) = f_{xx^i}(\ldots) + (B_o)_{x^i}(\ldots)v(t)^T + v^i(t)(B_o)_x(\ldots). \end{aligned}$$

The first relation leads immediately to a representation for $\hat{A}(\cdot)$, namely

$$(23.37)\qquad \hat{A}(t) = A(t) + B_o(t)v(t)^T$$

(for the definition of $B_o(\cdot)$ cf. (23.29)). Since $y(\cdot)$ is orthogonal to $B_o(\cdot)$ (first order necessary condition!) it is clear that $y(\cdot)$ is also a solution of the linear differential equation

$$\dot{y} = -\hat{A}^T(t)y \ .$$

Therefore $y(\cdot)$ and $\hat{y}(\cdot)$ coincide, because they have the same initial value at $t=t_e$. Furthermore one finds from the second of the relations (23.36) that

$$y(t)^T\hat{f}_{x^ix^j}(\ldots) = y(t)^T f_{x^ix^j}(\ldots)+y(t)^T(B_o)_{x^i}(\ldots)v^j(t)$$
$$+ y(t)^T(B_o)_{x^j}(\ldots)v^i(t) \ .$$

Hence the matrix $\hat{K}(t,y(t))$ which forms the inhomogeneous part in the differential equation for $\hat{P}$ can be represented in the form

$$K(t,y(t)) + (B_o)_x(t,x(t))^T y(t)v(t)^T + v(t)y(t)^T(B_o)_x(t,x(t))$$

where $K(t,y(t))$ is the inhomogeneous part in the differential equation for $\tilde{P}$ (cf. (23.9)). Note that the additional term can also be written as

$$s(t)v(t)^T + v(t)s(t)^T$$

(cf. (23.29)). It is now clear, in view of the explicit formulas for $\hat{K}$ and $\hat{A}$ (cf. (23.37)) that the matrices $\hat{P}(\cdot)$ and $P(\cdot)$ (the latter one was introduced at the beginning of the proof) are solutions of the same differential equation (23.28). Since the values of $\hat{P}$ and P at $t=t_e$ are equal to the Hessian matrix of φ (for $x=\tilde{x}(t_e)$) it is clear that we have $\hat{P}(t)=P(t)$ for all t. The inequality which we are aiming at can therefore be written as

(23.30′) $\quad \tilde{B}^T\tilde{P}(\tilde{t})\tilde{B} \leq \tilde{B}^T\hat{P}(\tilde{t})\tilde{B}$.

In order to prove (23.30′) we make use of the fact that $v(\cdot)$ vanishes for all t in a neighborhood of $\tilde{t}$. f and $\hat{f}$ therefore coincide for all (t,x,u) provided t is sufficiently close to $\tilde{t}$. Hence one can find an open interval $\tilde{\mathcal{S}}$ which contains $\tilde{t}$ such that the computation of the B_ν (regarded as functions of $t,x,\boldsymbol{u}$) for the modified equation (23.35) yields the original B_ν, at least if t is restricted to $\tilde{\mathcal{S}}$. This observation provides the key for the proof. If the set $\mathcal{S}$ is identified with $\tilde{\mathcal{S}}$ all quantities - except $P(t)$ - which appear in the hypothesis and the conclusion of Theorem 23.1 are the same for the original and the modified equation (23.35). Hence for all $t\in\tilde{\mathcal{S}}$ the inequality

(23.38) $$\sum_{\mu,\nu=0}^{\rho} \xi_\mu\xi_\nu y^T(B_\mu)_x B_\nu + \Big(\sum_{\mu=0}^{\rho}\xi_\mu B_\mu^T\Big)P\Big(\sum_{\mu=0}^{\rho}\xi_\mu B_\mu\Big)\geq 0$$

holds both for $P = \tilde{P}(t)$ and $P = \hat{P}(t)$.

Finally the decisive hypothesis of the theorem comes into play: The above inequality turns into an equality if t,ξ_μ,P are specified as follows

$$t \to \tilde{t}, \xi_\mu \to \tilde{\xi}_\mu, \mu=0,\ldots,\rho, \; P \to \tilde{P}(\tilde{t}).$$

This means that the value of the expression (23.38) for $P=\tilde{P}(\tilde{t})$ is not bigger than the value for $P=P(t)$, i.e. (23.30′) holds true. Thereby the theorem is proved.

There exists a slightly strengthened version of the Jacobson-Gabosov condition which has been discussed by Warga [27]. This condition has a somehow global character in so far as it involves simultaneously values over the entire "singular regime". We wish to comment briefly upon it. Let us assume that the hypotheses of Theorem 23.1 are satisfy for some $\rho \geq 0$ and let us choose finitely many $t \in \mathcal{S}$, say $t_1 < t_2 < \ldots < t_k$. Let $\xi_{i,\mu}$, $i=1,\ldots,k, \mu=0,\ldots,\rho$, be independent variables. One can then find a quadratic form in these variables which is positive-semidefinite and has coefficients depending upon the values of the reference trajectory and the solutions of the equations (23.9) at $t=t_i, i=1,\ldots,k$. If all $\xi_{i,\mu}$ for $i \neq j$ are zero then the form reduces to (23.19) (for $\xi_\mu = \xi_{j,\mu}$). All this can be put into evidence by an obvious modification of the construction employed in the proof of Theorem 23.1. Instead of a single linear form $p(x)$ we consider simultaneously the k linear forms

$$p_i(x) := \sum_\mu \xi_{i,\mu} B_\mu(t_i, x). \quad \text{(cf. (23.23))}.$$

To each p_i corresponds a control variation concentrated at $t=t_i$ and depending upon $\xi_{i,\mu}$ and the parameter λ (λ is assumed to be independent from i). The superposition of these control variations generates then a family of admissible trajectories depending upon t, λ, and all $\xi_{i,\mu}$ which tends to the reference trajectory for $\lambda \to 0$. The conclusion follows now by standard arguments (cf. the corollary to Lemma 23.1).

Superposition of control variations have actually occured in the proof of Theorem 9.1 (Sec. 10,11). That they did not lead to tests other than those involving a single point t has a simple reason: Theorem 9.1 is a multiplier rule and hence based essentially on linear approximations which are additive under superposition.

APPENDIX

The first lemma is concerned with the independence of certain linear forms. The variables which appear in these forms are denoted by $v_{i,j}, i,j = 1,2,\ldots,$ each $v_{i,j}$ being a m-dimensional vector. The forms are m-dimensional vectors themselves and are given in explicit terms as follows

$$(1) \qquad s_{\nu,\rho} = \sum_{i=1}^{N} z_i^{\nu}(v_{i-1,\rho}-v_{i,\rho}) \ , \ \nu \geq 0, \ \rho \geq 0$$

with $v_{o,\rho} = 0$ for all ρ .

Lemma 1. Let N be a positive integer and $z_1,\ldots,z_N$ N different and non-vanishing real numbers. Then the components of $s_{\nu,\rho}$ for $\nu = \nu_o, \nu_o+1,\ldots\ldots,\nu_o+N-1, \rho = 0,1,2,\ldots\ldots$, form a set of independent scalar linear forms.

Proof. Let us assume that we have a relation

$$(2) \qquad \sum_{\rho} \sum_{\nu=\nu_o}^{\nu_o+N-1} k^T_{\nu,\rho} \ s_{\nu,\rho} = 0, \quad \text{identically in } v_{i,j} \ ,$$

where ρ runs through a finite set of different integers and the $k^T_{\nu,\rho}$ are some constant row vectors. We have to show that (2) implies that all $k^T_{\nu,\rho}$ vanish. Since the $v_{i-1,\rho}-v_{i,\rho}$ can be regarded as independent variables we obtain from (1) , (2) the relations

$$\sum_{\nu=\nu_o}^{\nu_o+N-1} z_i^{\nu} k^T_{\nu,\rho} = 0, \ i=1,\ldots,N \ ,$$

which hold for each ρ . Considered componentwise they constitute a set of N homogeneous equations for the respective components of $k^T_{\nu,\rho}$. The determinant of these equations is essentially the Vandermonde-determinant of $z_i, i = 1,\ldots, N$ and does not vanish. Therefore the linear equations in question admit the trivial solution only.

Corollary. The following linear forms in $v_{i,j}$ are independent: $s_{\nu,\rho}$ for $1 \leq \nu < N, \ \rho \geq 0,$ and $v_{N,\rho}, \rho \geq 0.$

Proof. Because of $v_{o,\rho} = 0$ we have $s_{o,\rho} = -v_{N,\rho}$.

In the sequel the notion "function" always refers to a scalar function of one variable t. Finitely many functions which are defined on some interval $[a,b]$ are called linearly independent with respect to $[a,b]$ if a linear combination with constant coefficients cannot vanish identically on $[a,b]$ unless all coefficients are zero. We make use of the following two elementary and well known statements concerning independency of funtions.

(i) Given continuous functions $\varphi_1(\cdot),\ldots,\varphi_N(\cdot)$, $\psi(\cdot)$ on some interval $[a,b]$. Assume that $\varphi_1(\cdot),\ldots,\varphi_N(\cdot)$ are linearly independent with respect to $[a,b]$ and that $\psi(t) \geq 0$ for all t. Furthermore assume that $\psi(t)$ does not vanish for all t in any open subset of $[a,b]$. Then the symmetric Matrix

$$\left(\int_a^b \psi(t)\varphi_i(t)\varphi_j(t)dt \right)$$

is positiv-definite.

(ii) Let $\Psi(\cdot),\varphi_i(\cdot)$ be as above and let there be given arbitrary real numbers $\alpha_1,\ldots,\alpha_N$. Then one can find a continuous function $\sigma(\cdot)$, which depends upon $\alpha_1,\ldots,\alpha_N$ such that

$$\int_a^b \psi(t)\sigma(t)\varphi_i(t)dt = \alpha_i, \quad i=1,\ldots,N. \tag{3}$$

Actually there exists a unique $\sigma(\cdot)$ which can be represented as a linear combination of $\varphi_1(\cdot),\ldots,\varphi_N(\cdot)$.

<u>Lemma 2</u> Given a positive integer K. Claim: One can find functions $z(\cdot),\xi(\cdot)$ which are of class C^2 on $\mathbb{R}$ and satisfy these conditions.

(i) $z(\cdot)$ is strictly increasing and negative for all t,

(ii) $\int_0^1 \xi(t)z(t)^\nu dt = 0$ for $\nu=0,\ldots,K$.

(iii) the functions

(4) $z(t)^\mu$, $z(t)^\mu\xi(t)$, $z(t)^\mu\int_0^t \xi(\tau)z(\tau)^\rho d\tau$, $\mu,\rho = 0,\ldots,K$

are linearly independent with respect to $[0,1]$.

<u>Proof.</u> It suffices to construct functions $z(\cdot)$, $\xi(\cdot)$ which have properties (i), (iii). Indeed, if $\xi(\cdot), z(\cdot)$ satisfy the first and third condition, then the $2K+2+(K+1)^2$ functions which appear in (4) are linearly independent with respect to some subinterval $[0,a]$ $a<1$. They remain linearly independent with respect to this interval and hence with respect to $[0,1]$ if $\xi(\cdot)$ is modified somehow on $(a,1]$. So we may assume without loss of generality that $\xi(\cdot)$ has a third order zero at some $\tilde{t} \in (a,1)$. Since the functions $z(t)^{\nu}, \nu = 0,\ldots$ are linearly independent with respect to any interval of positive length we can refer to the remarks concerning the integral equations (3): There exists a linear combination $\sigma(t)$ of the powers $z(t)^{\nu}$ with constant coefficients such that the relations

$$\int_0^{\tilde{t}} \xi(t)z(t)^{\nu}dt + \int_{\tilde{t}}^1 (t-\tilde{t})^3\sigma(t)z(t)^{\nu}dt = 0, \quad \nu=0,\ldots,K$$

hold. Hence if we define $\xi(\cdot)$ an $[0,\tilde{t}]$ as before and if we put

$$\xi(t) = (t-\tilde{t})^3\sigma(t) \quad \text{for} \quad t > \tilde{t}$$

then clearly (ii) and (iii) hold simultaneously.

Let us now put $z(t) = -e^{-t}$ and take as $\xi(\cdot)$ an entire function of exponential type. All functions appearing in (4) are then also entire and of exponential type. Hence if they are linearly dependent with respect to $[0,1]$ they are also linearly dependent with respect to $[0,\infty)$ and so are their Laplace transforms (with respect to the field of complex numbers). Let $\hat{\xi}(s)$ be the Laplace transform of $\xi(t)$. Then, by standard rules, the Laplace transforms of the functions (4) are

$$(5) \quad \frac{1}{s+\mu}\ , \quad \hat{\xi}(s+\mu)\ , \quad \frac{1}{s+\mu}\hat{\xi}(s+\rho+\mu), \quad \mu,\rho = 0,\ldots,K$$

except for a factor ± 1. Linear dependence of the functions (5) implies that $\hat{\xi}$ satisfies a linear difference equation with rational functions as coefficients. Hence the Laplace transforms (5) cannot be linearly dependent if $\hat{\xi}$ has a finite number of essential singularities. A typical example of such a function is $\xi(s) = (1/s)\exp(-1/s)$ which indeed is known to be the Laplace transform of an entire function of exponential type, namely

$$\xi(t) = J_o(2\sqrt{t}),$$

$J_o(\cdot)$ being the Bessel function of the first kind of order zero. Thereby the lemma is proved.

Lemma 3. Given a positive integer K. Claim: One can find functions $\eta(\cdot)$, $z(\cdot)$ which are of class C^2 on $[0,1]$ and satisfy these conditions.

(i) $z(t) < 0$ for all t, $z(\cdot)$ is strictly increasing, $\eta(0) = 0$,

(ii) $\int_0^1 \dot{\eta}(t)z(t)^\nu dt = 0, \quad \nu = 0,1,\dots,K,$

(iii) Given arbitrary real numbers $a_i, b_i, c_i, d_i, e_{\mu,\nu}$, $0 \le i \le K$, $0 < \mu, \nu \le K$. Then there exists a function $\rho(t)$ which is of class C^2 on $[0,1]$ such that $\rho(0) = 0$ and the following relations hold

1) $\int_0^1 \dot{\rho}(t)z(t)^\nu dt = a_\nu, \quad \nu=0,\dots,K,$

2) $\int_0^1 \dot{\rho}(t)\eta(t)z(t)^\nu dt = b_\nu, \quad \nu= 0,\dots,K,$

3) $\int_0^1 \rho(t)\dot{\eta}(t)z(t)^\nu dt = c_\nu, \quad \nu= 0,\dots,K,$

4) $\int_0^1 \dot{\rho}(t)\dot{\eta}(t)z(t)^\nu dt = d_\nu, \quad \nu= 0,\dots,K,$

5) $\int_0^1 \dot{\rho}(t)z(t)^\mu \int_0^t \dot{\eta}(\tau)z(\tau)^\nu d\tau dt = e_{\mu,\nu} \quad 0< \mu,\nu \le K.$

Remark. $\rho(\cdot)$ depends upon $a_\nu, b_\nu, c_\nu, d_\nu, e_{\nu,\mu}$. Actually one can always find a ρ which is a linear function of these quantities as will become apparent from the proof.

Proof. We first remark that the conditions 2) and 3) respectively can be replaced by

$$2')\quad \int_0^1 \dot\rho(t)z(t)^\nu \int_0^t \dot\eta(\tau)d\tau \qquad \nu=0,\dots,K,$$

$$3')\quad -\int_0^1 \dot\rho(t) \int_0^t \dot\eta(\tau)z(\tau)^\nu\, d\tau dt = c_\nu,\ \nu=0,\dots,K,$$

respectively. This is clear - because of $\eta(0) = 0$- in case of 2'). It follows by integration by parts in case 3') if one takes condition (ii) into account. 2') and 3') now can be formally subsumed among 5) if one admits $\mu=0$, $\nu=0$. Hence we can reformulate the problem as follows: Find $\rho(\cdot)$, $\eta(\cdot)$, $z(\cdot)$ such that (i), (ii) hold and the following set of relations is satisfied

$$\int_0^1 \dot\rho(t)z(t)^\nu dt = a_\nu ,$$

(6)
$$\int_0^1 \dot\rho(t)\dot\eta(t)z(t)^\nu dt = d_\nu ,$$

$$\int_0^1 \dot\rho(t)z(t)^\mu \int_0^t \dot\eta(\tau)z(\tau)^\nu d\tau dt = e_{\mu,\nu} .$$

Here ν,μ vary independently from 0 to K, a_ν, d_ν, $e_{\nu,\mu}$ are arbitrary real numbers.

We now choose $z(\cdot)$, $\xi(\cdot)$ as in Lemma 2 and define

$$\eta(t) := \int_0^t \xi(\tau)d\tau.$$

(6) then is nothing else than a system of integral equations of the form (3), $\dot\rho(\cdot)$ is the unknown function, $\psi(t) \equiv 1$ and the role of the φ_i is played by the functions appearing in (4). The latter ones are by construction linearly independent with respect to $[0,1]$. Hence the system can be solved in terms of a function $\sigma(\cdot)$ which can be written as a linear combination of the functions (4). If we put

$$\rho(t) := \int_0^t \sigma(\tau)d\tau$$

we have found a function having the desired properties.

Our aim is to establish a statement analogous to Lemma 3, where the integrals are replaced by finite sums (Lemma 5). As a preparation we need

Lemma 4. Given functions $\xi(\cdot)$, $\eta(\cdot)$, $\zeta(\cdot)$, $\kappa(\cdot)$ which are of class C^2 on $[0,1]$. Let N be a positive integer and let $S_\nu(N)$ be given as follows:

$$S_1(N) := \sum_{i=1}^{N} (\xi((i-1)/N) - \xi(i/N))\zeta(t_i) ,$$

$$S_2(N) := \sum_{i=1}^{N} (\xi((i-1)/N) - \xi(i/N))(\eta((i-1)/N)-\eta(i/N))\zeta(t_i) ,$$

$$S_3(N) := \sum_{1\le i<j\le N} (\xi((i-1)/N)-\xi(i/N))(\eta((j-1)/N)-\eta(j/N))\zeta(t_i)\kappa(t_j')$$

where $(i-1)/N \le t_i, t_i' \le i/N$, $i=1,\dots,N$.

Claim: For $N\to\infty$ these asymptotic formulas hold true

$$S_1(N) = -\int_0^1 \dot{\xi}(t)\zeta(t)dt + \mathcal{O}(\tfrac{1}{N}),$$

$$S_2(N) = \frac{1}{N}\int_0^1 \dot{\xi}(t)\dot{\eta}(t)\zeta(t)dt + \mathcal{O}(\tfrac{1}{N^2}) ,$$

$$S_3(N) = \int_0^1 \dot{\xi}(t)\zeta(t)\int_t^1 \dot{\eta}(\tau)\kappa(\tau)d\tau dt + \mathcal{O}(\tfrac{1}{N}) .$$

Furthermore one can find explicit estimates for the $\mathcal{O}$-expressions in terms of upper bounds (on the interval $[0,1]$) for the absolute value of the functions $\xi(\cdot)$, $\eta(\cdot)$, $\zeta(\cdot)$, $\kappa(\cdot)$ and their first and second order derivatives.

Proof. The proof is straightforward, using the definition of Riemann integral and standard mean value theorems.

Lemma 5. Given a positive integer K. Claim: There exists a positive integer M, real numbers $z_1,\dots,z_M$, a M-tuple $\underline{\eta} = (\eta_1,\dots,\eta_M)$ of real numbers such that the subsequent statements (i)-(iii) are true (with $\eta_o = 0$).

(i) $z_1 < z_2 < \dots < z_M < 0$,

(ii) $\mathcal{L}^{(\nu)}(\underline{\eta}) := \sum_{i=1}^{M} (\eta_{i-1}-\eta_i)z_i^\nu = 0$, $\nu = 0,\dots,K$.

(iii) the following linear forms in the variables $\xi_1, \ldots, \xi_M$ are linearly independent (we put $\xi_o = 0$ and write $\underline{\xi}$ for the M-tuple $(\xi_1, \ldots, \xi_M)^T$) :

$$\mathcal{L}_M^{(\nu)}(\underline{\xi}) := \sum_{i=1}^{M} (\xi_{i-1}-\xi_i) z_i^{\nu} , \quad \nu = 0, \ldots, K ,$$

$$S_M^{(\nu)}(\underline{\eta}, \underline{\xi}) := \sum_{i=1}^{M} (\eta_{i-1}-\eta_i)(\xi_{i-1}+\xi_i) z_i^{\nu} , \quad \nu = 0, \ldots, K,$$

$$S_M^{(\nu)}(\underline{\xi}, \underline{\eta}) := \sum_{i=1}^{M} (\xi_{i-1}-\xi_i)(\eta_{i-1}+\eta_i) z_i^{\nu} , \quad \nu = 0, \ldots, K,$$

$$P_M^{(\nu)}(\underline{\xi}, \underline{\eta}) := \sum_{i=1}^{M} (\xi_{i-1}-\xi_i)(\eta_{i-1}-\eta_i) z_i^{\nu} , \quad \nu = 0, \ldots, K,$$

$$Q_M^{(\tau,\nu)}(\underline{\eta}, \underline{\xi}) := \sum_{1 \leq i < j \leq M} (\eta_{i-1}-\eta_i)(\xi_{j-1}-\xi_j) z_i^{\tau} z_j^{\nu} , \quad \nu, \tau = 1, \ldots, K.$$

Remark. Because of the identity (21.11) one could replace the $Q^{(\tau,\nu)}(\underline{\eta}, \underline{\xi})$ by the linear forms $Q^{(\nu,\tau)}(\underline{\xi}, \underline{\eta})$ without affecting the statement of the lemma. The linear forms so obtained are precisely the ones which appear in (21.7), (21.8).

Proof. We first note that (N.B. $\xi_o = 0$!)

$$\mathcal{L}^{(o)}(\underline{\xi}) = \xi_M . \tag{7}$$

Next we wish to convince ourselves that the lemma is proved once we have found M, z_i, η_i such that conditions (i), (iii) are satisfied. Indeed, let us assume that (i) holds and that the linear forms in $\xi_1, \ldots, \xi_M$ which appear in (iii) are independent. Choose then further real numbers $z_{M+1}, \ldots, z_{M+K+1}$ such that

$$z_M < z_{M+1} < \ldots < z_{M+K+1} < 0$$

and solve the $K+1$ linear equations

$$\sum_{i=1}^{M} (\eta_{i-1}-\eta_i) z_i^{\nu} + \sum_{i=M+1}^{M+K+1} (\eta_{i-1}-\eta_i) z_i^{\nu} = 0, \quad \nu=0, \ldots, K$$

for $\eta_{M+1},\ldots,\eta_{M+k+1}$. Consider now the linear forms in $\xi_1,\ldots\ldots,\xi_M,\ \xi_{M+1},\ldots,\ \xi_{M+K+1}$, which are associated with the enlarged system of z_i,η_i. We claim that these forms are also independent. Otherwise they would turn into a system of dependent linear forms in $\xi_1,\ldots,\xi_M$ if one substitutes ξ_M for those ξ_i where $i > M$. That this cannot be true can be put into evidence by means of a simple observation: As a result of the substitution $\xi_i \to \xi_M$, $i > M$, one obtains linear forms which differ from the ones listed in (iii) by terms which are multiples of ξ_M. Because of (7) it is then clear that this new system actually is equivalent to the original one, therefore it consists of independent forms.

In order to complete the proof it remains the problem to find M, z_i, η_i such that (i) holds and the linear forms (iii) are independent. We choose functions $z(t)$, $\eta(t)$ as described in Lemma 3. Next we divide the interval $[0,1]$ in M equal parts and put $\eta_i = \eta(i/M)$, $z_i = z(i/M)$. Our aim is to show that for this choice of η_i, z_i the linear forms in question become independent if M is sufficiently large. Let us assume that the contrary is true. In other words assume that we have for every M a non-trivial relation between the linear forms appearing in hypothesis (iii). In order to write down this relation in a convenient form let us denote by $\mathcal{L}_M(\underline{\xi}), S_M^{(\nu)}(\underline{\eta},\underline{\xi})$ etc. the column vector with the components $\mathcal{L}_M^{(\nu)}(\underline{\xi})$, $S_M^{(\nu)}(\underline{\eta},\underline{\xi})$ etc.. The assumption can then be phrased as follows: For every M there exists column vectors l_M, s_M, $\tilde{s}_M, p_M, q_M$ of appropriate dimensions and not all vanishing such that

$$(8)\quad l_M^T\,\mathcal{L}_M(\underline{\xi})+s_M^T S_M(\underline{\eta},\underline{\xi})+\tilde{s}_M^T S_M(\underline{\xi},\underline{\eta})+p_M^T P_M(\underline{\xi},\underline{\eta})+q_M^T Q_M(\underline{\eta},\underline{\xi}) = 0 .$$

Without loss of generality we may assume that

$$(9)\quad \| l_M \|^2 + \| s_M \|^2 + \| \tilde{s}_M \|^2 + \| p_M \|^2 + \| q_M \|^2 = 1 .$$

The next step of the proof will be performed with the help of Lemma 3 and the subsequent remark. We wish to represent the components of l_M, s_M etc. as definite integrals of the particular type considered previously. To this purpose we choose a sequence of functions $\rho_M(\cdot)$ which are of class C^2 on $[0,1]$ and which satisfy the following conditions.

(i) $\rho_M(\cdot), \dot\rho_M(\cdot), \ddot\rho_M(\cdot)$ is bounded on $[0,1]$ independently from M,

(ii). the ν-th component of $l_M, s_M, \tilde{s}_M, p_M$ respectively is given by the integral

$$-\int_0^1 \dot\rho_M z^\nu dt, \quad -2\int_0^1 \dot\rho_M \eta z^\nu dt, \quad -2\int_0^1 \rho_M \dot\eta z^\nu dt, \quad \int_0^1 \dot\rho_M \dot\eta z^\nu dt$$

(10) respectively (we have omitted the argument t in $\rho_M, \dot\rho_M, z, \eta, \dot\eta$),

(iii) the real number which occupies the place (ν,μ) in the vector q_M equals the integral

$$\int_0^1 \dot\rho_M(t) z(t)^\mu \int_0^t \dot\eta(\tau) z(\tau)^\nu d\tau dt.$$

We now are in the position to demonstrate that for sufficiently large M the two relations (8) and (9) are not compatible. This is done, roughly speaking, by specializing $\underline{\xi}$ in such a way that $\mathcal{L}_M(\underline{\xi}) \approx l_M$, $S_M(\underline{\eta},\underline{\xi}) \approx s_M$, $S_M(\underline{\xi},\underline{\eta}) \approx \tilde{s}_M$, $P_M(\underline{\xi},\underline{\mu}) \approx p_M$, $Q_M(\underline{\eta},\underline{\xi}) \approx q_M$. In order to make this argument precise we refer to Lemma 4 and construct approximations to the integrals listed under (10). It is not difficult to verify that the leading terms in these approximations coincide with the sums listed under hypothesis (iii) of Lemma 5 if ξ_i is specialized to the real number $\rho_M(i/M), i=1,\dots,M$. Hence the assertion of Lemma 4 and the statement (9) combined yield the following asymptotic formula for $M\to\infty$ (where $\underline{\xi}$, for every fixed M, has to be specialized as indicated above):

$$(\mathcal{L}_M(\underline{\xi}), S_M(\underline{\eta},\underline{\xi}), S_M(\underline{\xi},\underline{\eta}), Q_M(\underline{\eta},\underline{\xi})) = r_M + \mathcal{O}(\tfrac{1}{M}), \quad P_M(\underline{\xi},\underline{\eta}) = \tfrac{1}{M} p_M + \mathcal{O}(\tfrac{1}{M^2}).$$

Here we have denoted by r_M the 4-tuple $(l_M, s_M, \tilde{s}_M, q_M)$. From (8) and (9) we now have the asymptotic relations for $M\to\infty$

$$\| r_M \|^2 + \tfrac{1}{M} \| p_M \|^2 = \tfrac{1}{M} \mathcal{O}(\| r_M \|) + \tfrac{1}{M^2} \mathcal{O}(\| p_M \|)$$

$$\| r_M \|^2 + \| p_M \|^2 = 1$$

which clearly contradict each other. Note that the first relation implies

$$\| r_M \|^2 = \mathcal{O}(\tfrac{1}{M}), \quad \tfrac{1}{M}\| p_M \|^2 = \tfrac{1}{M} \mathcal{O}(\| r_M \|) + \tfrac{1}{M^2} \mathcal{O}(\| p_M \|).$$

REFERENCES

[1] H.W.Knobloch and F.Kappel, Gewöhnliche Differentialgleichungen B.G.Teubner, Stuttgart 1974.

[2] A.J.Krener, "The high order maximal principle", in Geometric Methods in System Theory, D.Q.Mayne and R.W.Brockett eds., D.Reidel, Dordrecht, Holland 1973, pp. 174-184.

[3] A.J.Krener, "The high order maximal principle and its application to singular extremals", SIAM J.Control Optimization 15 (1977), pp. 256-293.

[4] R.W.Brockett, "Lie theory, functional expansions, and necessary conditions" in Mathematical Control Theory, Lecture Notes in Mathematics Vol. 680, Springer-Verlag Berlin, Heidelberg, New York 1978, pp. 68-76.

[5] R.W.Brockett, "Functional expansions and higher order necessary conditions in optimal control" in Mathematical Systems Theory, Lecture Notes in Economics and Mathematical Systems Vol. 131, Springer-Verlag Berlin, Heidelberg, New York 1976, pp. 111-121.

[6] D.J.Clements and B.D.O.Anderson, Singular Optimal Control: The Linear-Quadratic Problem. Lecture Notes in Control and Information Sciences Vol. 5, Springer-Verlag, Berlin, Heidelberg, New York 1978.

[7] D.J.Bell and D.H.Jacobson, Singular optimal control problems, Academic Press, London 1975.

[8] H.J.Kelley, R.Kopp and H.G.Moyer, "Singular Extremals". in: Topics in Optimization (G.Leitmann ed.), Academic Press, New York and London 1967, Chapter 3.

[9] B.S.Goh, "Necessary conditions for singular extremals involving multiple control variables", SIAM J.Control 5 (1966),pp.716-731.

[10] M.R.Hestenes, Calculus of Variations and Optimal Control Theory, John Wiley, New York-London-Sidney 1966.

[11] R.Gabasov and F.Kirillova,The Qualitative Theory of Optimal Processes, Marcel Dekker Inc., New York and Basel 1976.

[12] D.H.Jacobson, "A new necessary condition of optimality for singular control problems", SIAM J.Control 7(1969),pp. 578-595.

[13] H.J.Kelley, "A second variation test for singular extremals", AIAA J.Vol.2 (1964), pp. 1380-1382.

[14] D.H.Jacobson and J.L.Speyer, "Necessary and sufficient conditions for optimality for singular control problems: A limit approach." J.Math.Anal.Appl.34 (1971),pp. 239-266.

[15] H.J.Kelley, "A transformation approach to singular subarcs in optimal trajectory and control problems", SIAM J.Control 2 (1964), pp. 234-240.

[16] J.V.Breakwell, "A doubly singular problem in optimal interplanetary guidance", SIAM J. Control 3 (1965), pp. 71-77.

[17] R.Gabasov and F.M.Kirillova, "High order necessary conditions for optimality", SIAM J.Control 10 (1972), pp. 127-168.

[18] J.L.Speyer and D.H.Jacobson, "Necessary and sufficient conditions for optimality for singular control problems: A transformation approach", J.Math.Anal.Appl.33 (1971),pp.163-187.

[19] H.J.Sussmann and V.Jurdjevic, "Controllability of nonlinear systems", J.Diff.Eqs. 12 (1972),pp. 95-116.

[20] H.Hermes and G.W.Haynes, "On the nonlinear control problem with control appearing linearly", J. SIAM Control, 1 (1963),pp. 85-108.

[21] H.Hermes, "On local and global control controllability" SIAM J.Control Optimization 12 (1974),pp.252-261.

[22] H.Hermes, "Local controllability and sufficient conditions in singular problems", J.Diff.Eqs. 20 (1976), pp.213-232.

[23] H.M. Robbins, "Optimality of intermediate thrust arcs of rocket trajectories", AIAA J.Vol.3 (1965), pp.1094-1098.

[24] R.Gabasov and F.M.Kirillova, "High order necessary conditions for optimality", SIAM J.Control Optimization 10 (1972),pp.127-168.

[25] H.J.Sussmann, "A sufficient condition to local controllability", SIAM J.Control Optimization 16 (1978),pp. 790-802.

[26] R.M.Lewis, "Definitions of order and junction conditions in singular optimal control problems" SIAM J.Control Optimization 18 (1980), pp. 21-32.

[27] J.Warga, "A second-order condition that strengthens Pontryagin's maximum principle", J.Diff.Eqs. 28 (1978),pp.284-307.

[28] C.Vârsan, "Higher order local controllability", Revue Roum. Math. Pures Appl.12(1975) pp. 103-118.

[29] C.Vârsan, "Local controllability along a singular trajectory," Revue Roum. Math. Pures Appl. 22 (1977), pp. 1011-1020.

[30] R.Gamkrelidze, "Exponential representations of solutions of ordinary differential equations" in <u>Equadiff IV</u>, Lecture Notes in Mathematics Vol. 703, Springer Verlag Berlin, Heidelberg, New York 1979, pp.110-129.

Lecture Notes in Control and Information Sciences

Edited by A. V. Balakrishnan and M. Thoma

Vol. 22: Optimization Techniques
Proceedings of the 9th IFIP Conference on
Optimization Techniques,
Warsaw, September 4–8, 1979
Part 1
Edited by K. Iracki, K. Malanowski, S. Walukiewicz
XVI, 569 pages. 1980

Vol. 23: Optimization Techniques
Proceedings of the 9th IFIP Conference on
Optimization Techniques,
Warsaw, September 4-8, 1979
Part 2
Edited by K. Iracki, K. Malanowski, S. Walukiewicz
XV, 621 pages. 1980

Vol. 24: Methods and Applications
in Adaptive Control
Proceedings of an International Symposium
Bochum, 1980
Edited by H. Unbehauen
VI, 309 pages. 1980

Vol. 25: Stochastic Differential Systems –
Filtering and Control
Proceedings of the IFIP-WG7/1 Working Conference
Vilnius, Lithuania, USSR, Aug. 28 – Sept. 2, 1978
Edited by B. Grigelionis
X, 362 pages. 1980

Vol. 26: D. L. Iglehart, G. S. Shedler
Regenerative Simulation of Response
Times in Networks of Queues
XII, 204 pages. 1980

Vol. 27: D. H. Jacobson, D. H. Martin, M. Pachter, T. Geveci
Extensions of Linear-Quadratic Control Theory
XI, 288 pages. 1980

Vol. 28: Analysis and Optimization of Systems
Proceedings of the Fourth International
Conference on Analysis and Optimization of Systems
Versailles, December 16–19, 1980
Edited by A. Bensoussan and J. L. Lions
XIV, 999 pages. 1980

Vol. 29: M. Vidyasagar,
Input-Output Analysis of Large-Scale
Interconnected Systems –
Decomposition, Well-Posedness and Stability
VI, 221 pages. 1981

Vol. 30: Optimization and Optimal Control
Proceedings of a Conference Held at
Oberwolfach, March 16–22, 1980
Edited by A. Auslender, W. Oettli, and J. Stoer
VIII, 254 pages. 1981

Vol. 31: Berc Rustem
Projection Methods in Constrained
Optimisation and Applications
to Optimal Policy Decisions
XV, 315 pages. 1981

Vol. 32: Tsuyoshi Matsuo,
Realization Theory of
Continuous-Time Dynamical Systems
VI, 329 pages, 1981

Vol. 33: Peter Dransfield
Hydraulic Control Systems –
Design and Analysis of Their Dynamics
VII, 227 pages, 1981

Vol. 34: H.W. Knobloch
Higher Order Necessary
Conditions in Optimal
Control Theory
V, 173 pages, 1981